Integer Programming

**WILEY-INTERSCIENCE
SERIES IN DISCRETE MATHEMATICS AND OPTIMIZATION**

ADVISORY EDITORS

RONALD L. GRAHAM
AT & T Laboratories, Florham Park, New Jersey, U.S.A.

JAN KAREL LENSTRA
*Department of Mathematics and Computer Science,
Eindhoven University of Technology, Eindhoven, The Netherlands*

ROBERT E. TARJAN
*Princeton University, New Jersey, and
NEC Research Institute, Princeton, New Jersey, U.S.A.*

A complete list of titles in this series appears at the end of this volume.

Integer Programming

LAURENCE A. WOLSEY

A Wiley-Interscience Publication
JOHN WILEY & SONS, INC.
New York • Chichester • Weinheim • Brisbane • Singapore • Toronto

This text is printed on acid-free paper. ⊗

Copyright © 1998 by John Wiley & Sons, Inc. All rights reserved.

Published simultaneously in Canada.

No part of this publication may be reproduced, stored in a retrieval system or transmitted in any form or by any means, electronic, mechanical, photocopying, recording, scanning or otherwise, except as permitted under Sections 107 or 108 of the 1976 United States Copyright Act, without either the prior written permission of the Publisher, or authorization through payment of the appropriate per-copy fee to the Copyright Clearance Center, 222 Rosewood Drive, Danvers, MA 01923, (978) 750-8400, fax (978) 750-4744. Requests to the Publisher for permission should be addressed to the Permissions Department, John Wiley & Sons, Inc., 605 Third Avenue, New York, NY 10158-0012, (212) 850-6011, fax (212) 850-6008, E-Mail: PERMREQ@WILEY.COM.

Library of Congress Cataloging-in-Publication Data:

Wolsey, Laurence A.
    Integer programming / Laurence A. Wolsey
       p.   cm. — (Wiley-Interscience series in discrete mathematics and optimization)
    "Wiley-Interscience publication."
    Includes bibliographical references and indexes.
    ISBN 0-471-28366-5 (alk. paper)
    1. Integer programming. I. Title. II. Series.
T57.74W67   1998
519.7'7—dc21   98-7296

Printed in the United States of America

10 9 8 7 6 5 4 3 2 1

To Marguerite

Contents

Preface		xiii
Abbreviations and Notation		xvii
1	**Formulations**	**1**
	1.1 Introduction	1
	1.2 What Is an Integer Program?	3
	1.3 Formulating IPs and BIPs	5
	1.4 The Combinatorial Explosion	8
	1.5 Mixed Integer Formulations	9
	1.6 Alternative Formulations	12
	1.7 Good and Ideal Formulations	14
	1.8 Notes	18
	1.9 Exercises	19
2	**Optimality, Relaxation, and Bounds**	**23**
	2.1 Optimality and Relaxation	23
	2.2 Linear Programming Relaxations	25
	2.3 Combinatorial Relaxations	26
	2.4 Lagrangian Relaxation	27
	2.5 Duality	28

2.6	Primal Bounds: Greedy and Local Search	30
2.7	Notes	33
2.8	Exercises	33

3 Well-Solved Problems — 37
- 3.1 Properties of Easy Problems — 37
- 3.2 IPs with Totally Unimodular Matrices — 38
- 3.3 Minimum Cost Network Flows — 40
- 3.4 Special Minimum Cost Flows — 42
  - 3.4.1 Shortest Path — 42
  - 3.4.2 Maximum $s-t$ Flow — 43
- 3.5 Optimal Trees — 43
- 3.6 Submodularity and Matroids* — 46
- 3.7 Notes — 49
- 3.8 Exercises — 50

4 Matchings and Assignments *(cursory understanding / go / over)* — 53
- 4.1 Augmenting Paths and Optimality — 53
- 4.2 Bipartite Maximum Cardinality Matching — 55
- 4.3 The Assignment Problem — 57
- 4.4 Notes — 62
- 4.5 Exercises — 63

5 Dynamic Programming — 67
- 5.1 Some Motivation: Shortest Paths — 67
- 5.2 Uncapacitated Lot-Sizing — 68
- 5.3 An Optimal Subtree of a Tree — 71
- 5.4 Knapsack Problems — 72
  - 5.4.1 0–1 Knapsack — 73
  - 5.4.2 Integer Knapsack Problems — 74
- 5.5 Notes — 77
- 5.6 Exercises — 78

6 Complexity and Problem Reductions — 81
- 6.1 Complexity — 81
- 6.2 Decision Problems, and Classes $\mathcal{NP}$ and $\mathcal{P}$ — 82
- 6.3 Polynomial Reduction and the Class $\mathcal{NPC}$ — 84
- 6.4 Consequences of $\mathcal{P} = \mathcal{NP}$ or $\mathcal{P} \neq \mathcal{NP}$ — 87

	6.5	Optimization and Separation	88
	6.6	Notes	89
	6.7	Exercises	89
7	Branch and Bound		91
	7.1	Divide and Conquer	91
	7.2	Implicit Enumeration	92
	7.3	Branch and Bound: An Example	95
	7.4	LP-based Branch and Bound	98
	7.5	Using a Branch-and-Bound System	101
		7.5.1 If All Else Fails	103
	7.6	Preprocessing*	103
	7.7	Notes	107
	7.8	Exercises	108
8	Cutting Plane Algorithms		113
	8.1	Introduction	113
	8.2	Some Simple Valid Inequalities	114
	8.3	Valid Inequalities	117
		8.3.1 Valid Inequalities for Linear Programs	117
		8.3.2 Valid Inequalities for Integer Programs	118
	8.4	A Priori Addition of Constraints	121
	8.5	Automatic Reformulation or Cutting Plane Algorithms	123
	8.6	Gomory's Fractional Cutting Plane Algorithm	124
	8.7	Mixed Integer Cuts	127
		8.7.1 The Basic Mixed Integer Inequality	127
		8.7.2 The Mixed Integer Rounding (MIR) Inequality	129
		8.7.3 The Gomory Mixed Integer Cut*	129
	8.8	Disjunctive Inequalities*	130
	8.9	Notes	133
	8.10	Exercises	134
9	Strong Valid Inequalities		139
	9.1	Introduction	139
	9.2	Strong Inequalities	140
		9.2.1 Dominance	140
		9.2.2 Polyhedra, Faces, and Facets	142

		9.2.3 Facet and Convex Hull Proofs*	144
	9.3	0–1 Knapsack Inequalities	147
		9.3.1 Cover Inequalities	147
		9.3.2 Strengthening Cover Inequalities	148
		9.3.3 Separation for Cover Inequalities	150
	9.4	Mixed 0–1 Inequalities	151
		9.4.1 Flow Cover Inequalities	151
		9.4.2 Separation for Flow Cover Inequalities	153
	9.5	The Optimal Subtour Problem	154
		9.5.1 Separation for Generalized Subtour Constraints	155
	9.6	Branch-and-Cut	157
	9.7	Notes	160
	9.8	Exercises	161
10	Lagrangian Duality		167
	10.1	Lagrangian Relaxation	167
	10.2	The Strength of the Lagrangian Dual	172
	10.3	Solving the Lagrangian Dual	173
	10.4	Lagrangian Heuristics and Variable Fixing	177
	10.5	Choosing a Lagrangian Dual	179
	10.6	Notes	180
	10.7	Exercises	181
11	Column Generation Algorithms		185
	11.1	Introduction	185
	11.2	Dantzig-Wolfe Reformulation of an IP	187
	11.3	Solving the Master Linear Program	188
		11.3.1 STSP by Column Generation	190
		11.3.2 Strength of the Linear Programming Master	192
	11.4	IP Column Generation for 0–1 IP	193
	11.5	Implicit Partitioning/Packing Problems	194
	11.6	Partitioning with Identical Subsets*	196
	11.7	Notes	200
	11.8	Exercises	201
12	Heuristic Algorithms		203
	12.1	Introduction	203

12.2	Greedy and Local Search Revisited	204
12.3	Improved Local Search Heuristics	207
	12.3.1 Tabu Search	207
	12.3.2 Simulated Annealing	208
	12.3.3 Genetic Algorithms	210
12.4	Worst-Case Analysis of Heuristics	211
12.5	MIP-based Heuristics	214
12.6	Notes	217
12.7	Exercises	218

13 From Theory to Solutions

13.1	Introduction	221
13.2	Software for Solving Integer Programs	221
13.3	How Do We Find an Improved Formulation?	223
	13.3.1 Uncapacitated Lot-Sizing	223
	13.3.2 Capacitated Lot-Sizing	227
13.4	Fixed Charge Networks: Reformulations	229
	13.4.1 The Single Source Fixed Charge Network Flow Problem	229
	13.4.2 The Directed Subtree Problem	231
13.5	Multi-Item Single Machine Lot-Sizing	232
13.6	A Multiplexer Assignment Problem	236
13.7	Notes	240
13.8	Exercises	241

References 245

Index 261

Preface

Intended Audience

The book is addressed to undergraduates and graduates in operations research, mathematics, engineering, and computer science. It should be suitable for advanced undergraduate and Masters level programs. It is also aimed at users of integer programming who wish to understand why some problems are difficult to solve, how they can be reformulated so as to give better results, and how mixed integer programming systems can be used more effectively.

The book is essentially self-contained, though some familiarity with linear programming is assumed, and a few basic concepts from graph theory are used.

The book provides material for a one-semester course of 2–3 hours per week.

What and How

Integer Programming is about ways to solve optimization problems with discrete or integer variables. Such variables are used to model indivisibilities, and 0/1 variables are used to represent on/off decisions to buy, invest, hire, and so on. Such problems arise in all walks of life, whether in developing train or aircraft timetables, planning the work schedule of a production line or a maintenance team, or planning nationally or regionally the daily or weekly production of electricity.

The last ten years have seen a remarkable advance in our ability to solve to near optimality difficult practical integer programs. This is due to a combination of

i) Improved modeling

ii) Superior linear programming software

iii) Faster computers

iv) New cutting plane theory and algorithms

v) New heuristic methods

vi) Branch-and-cut and integer programming decomposition algorithms

Today many industrial users still build an integer programming model and stop at the first integer feasible solution provided by their software. Unless the problem is very easy, such solutions can be 5%, 10%, or 100% away from optimal, resulting in losses running into mega-dollars. In many cases it is now possible to obtain solutions that are proved to be optimal, or proven within 0.1%, 1%, or 5% of optimal, in a reasonable amount of computer time. There is, however, a cost: better models must be built, and either specially tailored algorithms must be constructed, or better use must be made of existing commercial software.

To make such gains, it is necessary to understand why some problems are more difficult than others, why some formulations are better than others, how effective different algorithms can be, and how integer programming software can be best used. The aim of this book is to provide some such understanding.

Chapter 1 introduces the reader to various integer programming problems and their formulation, and introduces the important distinction between good and bad formulations. Chapter 2 explains how it is possible to prove that feasible solutions are optimal or close to optimal.

Chapters 3–5 study integer programs that are easy. The problems and algorithms are interesting in their own right, but also because the algorithmic ideas can often be adapted so as to provide good feasible solutions for more difficult problems. In addition, these easy problems must often be solved repeatedly as subproblems in algorithms for the more difficult problems. We examine when linear programs automatically have integer solutions, which is in particular the case for network flow problems. The greedy algorithm for finding an optimal tree, the primal-dual algorithm for the assignment problem, and a variety of dynamic programming algorithms are presented, and their running times examined.

In Chapter 6 we informally address the question of the apparent difference in difficulty between the problems presented in Chapters 3–5 that can be solved rapidly, and the "difficult" problems treated in the rest of the book.

The fundamental branch-and-bound approach is presented in Chapter 7. Certain features of commercial integer programming systems based on branch-and-bound are discussed. In Chapters 8 and 9 we discuss valid inequalities and cutting planes. The use of inequalities to improve formulations and obtain tighter bounds is the area in which probably the most progress has been made in the last ten years. We give examples of the cuts, and also the routines to find cuts that are being added to the latest systems.

In Chapters 10 and 11 two important ways of decomposing integer programs are presented. The first is by Lagrangian relaxation and the second by column generation. It is often very easy to implement a special-purpose algorithm based on Lagrangian relaxation, and many applications are reported in the literature. Integer programming column generation, which is linear programming based, is more recent, but several recent applications suggest that its importance will grow.

Whereas the emphasis in Chapters 7–11 is on obtaining "dual" bounds (upper bounds on the optimal value of a maximization problem), the need to find good feasible solutions that provide "primal" (lower) bounds is addressed in Chapter 12. We present the basic ideas of various modern local search metaheuristics, introduce briefly the worst-case analysis of heuristics, and also discuss how an integer programming system can be used heuristically to find solutions of reasonable quality for highly intractible integer programs.

Finally, in Chapter 13 we change emphasis. By looking at a couple of applications and asking a series of typical questions, we try to give a better idea of how theory and practice converge when confronted with the choice of an appropriate algorithm, and the question of how to improve a formulation, or how to use a commercial mixed integer programming system effectively.

In using the book for a one-semester course, the chapters can be taken in order. In any case we suggest that the basics consisting of Chapters 1, 2, 3, 6, 7 should be studied in sequence. Chapter 4 is interesting for those who have had little exposure to combinatorial optimization. Chapter 5 can be postponed, and parts of Chapter 12 can be studied at any time after Chapter 2. There is also no difficulty in studying Chapter 10 before Chapters 8 and 9. The longer Chapters 7, 8, 9 and 11 contain starred sections that are optional. The instructor may wish to leave out some material from these chapters, or alternatively devote more time to them. Chapter 13 draws on material from most of the book, but can be used as motivation much earlier.

Acknowledgments

I am sincerely grateful to the many people who have contributed in some way to the preparation of this book. Marko Loparic has voluntarily played the role of teaching assistant in the course for which this material was developed. Michele Conforti, Cid de Souza, Eric Gourdin and Abilio Lucena have all used parts of it in the classroom and provided feedback. John Beasley, Marc Pirlot, Yves Pochet, James Tebboth and François Vanderbeck have criticized one or more chapters in detail, and Jan-Karel Lenstra has both encouraged and provided rapid feedback when requested. Finishing always takes longer than expected, and I am grateful to my colleagues and doctoral students at Core for their patience when they have tried to interest me in other more pressing matters, to the Computer Science Department of the University of Utrecht for allowing me to finish off the book in congenial and interesting surroundings, and to Esprit program 20118, MEMIPS, for support during part of the 1997–98 academic year. Sincere thanks go to Fabienne Henry for her secretarial help over many years, and for her work in producing the final manuscript.

Scientifically I am deeply indebted to the many researchers with whom I have had the good fortune and pleasure to collaborate. Working with George Nemhauser for many years has, I hope, taught me a little about writing. His earlier book on integer programming with R. Garfinkel provided an outstanding model for an undergraduate textbook. Yves Pochet is a considerate and stimulating colleague, and together with Bob Daniel, who has always been ready to provide a "practical problem per day," they provide a constant reminder that integer programming is challenging both theoretically and practically. However, the bias and faults that remain are entirely my own.

Abbreviations and Notation

BIP: Binary or Zero-One Integer Problem
B^n: $\{0,1\}^n$ the set of n-dimensional 0,1 vectors
C-G: Chvátal-Gomory
CLS: Capacitated Lot-Sizing Problem
conv(S): The convex hull of S
COP: Combinatorial Optimization Problem
D: Dual Problem
DP: Dynamic Programming
e_j: The jth unit vector
e^S: The characteristic vector of S
$E(S)$: All edges with both endpoints in the node set S
$FCNF$: Fixed Charge Network Flow Problem
GAP: Generalized Assignment Problem
GSEC: Generalized Subtour Elimination Constraint
GUB: Generalized Upper Bound
IKP: Integer Knapsack Problem
IP: Integer Programming Problem
IPM: Integer Programming Master Problem
LD: Lagrangian Dual Problem
lhs: left-hand side
LP: abbreviation for "linear programming"

LP: A specific or general Linear Programming Problem
LPM: Linear Programming Master Problem
L(X): Length of the input of a problem instance X
M: A large positive number
MIP: Mixed Integer Programming Problem
MIR: Mixed Integer Rounding
N: Generic set $\{1, 2, \cdots, n\}$
$\mathcal{NP}$: Class of NP problems
$\mathcal{NPC}$: Class of NP-complete problems
P: Generic Problem Class, or Polyhedron
$\mathcal{P}$: Class of polynomially solvable problems
$\mathcal{P}(N)$: Set of subsets of N
rhs: right-hand side
$RLPM$: Restricted Linear Programming Master Problem
R^n: The n-dimensional real numbers
R^n_+: The n-dimensional nonnegative real numbers
S: Feasible region of IP, or subset of N
SOS: Special Ordered Set
$STSP$: Symmetric Traveling Salesman Problem
TSP: (Asymmetric) Traveling Salesman Problem
TU: Totally Unimodular
UFL: Uncapacitated Facility Location Problem
ULS: Uncapacitated Lot-Sizing Problem
$V^-(i)$: Node set $\{k \in V : (k,i) \in A\}$
$V^+(i)$: Node set $\{k \in V : (i,k) \in A\}$
X: Feasible region of IP, or a problem instance
$(x)^+$: The maximum of x and 0
Z^n_+: The n-dimensional nonnegative integers
$\delta^-(S)$: All arcs going from a node not in S to a node in S
$\delta^+(S)$: All arcs going from a node in S to a node not in S
$\delta(S)$ or $\delta(S, V \setminus S)$: All edges with one endpoint in S and the other in $V \setminus S$
$\delta(i)$ or $\delta(\{i\})$: The set of edges incident to node i
1: The vector $(1, 1, \ldots, 1)$
0: The vector $(0, 0, \ldots, 0)$

Integer Programming

1
Formulations

1.1 INTRODUCTION

A wide variety of practical problems can be formulated and solved using integer programming. We start by describing briefly a few such problems.

1. Train Scheduling

Certain train schedules repeat every hour. For each line, the travel times between stations are known, and the time spent in a station must lie within a given time interval. Two trains traveling on the same line must for obvious reasons be separated by at least a given number of minutes. To make a connection between trains A and B at a particular station, the difference between the arrival time of A and the departure time of B must be sufficiently long to allow passengers to change, but sufficiently short so that the waiting time is not excessive. The problem is to find a feasible schedule.

2. Airline Crew Scheduling

Given the schedule of flights for a particular aircraft type, one problem is to design weekly schedules for the crews. Each day a crew must be assigned a duty period consisting of a set of one or more linking flights satisfying numerous constraints such as limited total flying time, minimum rests between flights, and so on. Then putting together the duty periods, weekly schedules or pairings are constructed which must satisfy further constraints on overnight rests, flying time, returning the crew to its starting point, and so on. The objective is to minimize the amount paid to the crews, which is a function of flying time, length of the duty periods and pairings, a guaranteed minimum number of flying hours, and so forth.

3. Production Planning

A multinational company holds a monthly planning meeting in which a three-month production and shipping plan is drawn up based on their latest estimates of potential sales. The plan covers 200–400 products produced in 5 different factories with shipments to 50 sales areas. Solutions must be generated on the spot, so only about 15 minutes' computation time is available. For each product there is a minimum production quantity, and production is in batches — multiples of some fixed amount. The goal is to maximize contribution.

4. Electricity Generation Planning

A universal problem is the unit commitment problem of developing an hourly schedule spanning a day or a week so as to decide which generators will be producing and at what levels. Constraints to be satisfied include satisfaction of estimated hourly or half-hourly demand, reserve constraints to ensure that the capacity of the active generators is sufficent should there be a sudden peak in demand, and ramping constraints to ensure that the rate of change of the output of a generator is not excessive. Generators have minimum on- and off-times, and their start-up costs are a nonlinear function of the time they have been idle.

5. Telecommunications

A typical problem given the explosion of demand in this area concerns the installation of new capacity so as to satisfy a predicted demand for data/voice transmission. Given estimates of the requirements between different centers, the existing capacity, and the costs of installing new capacity which is only available in discrete amounts, the problem is to minimize cost taking into account the possibility of failure of a line or a center due to a breakdown or accident.

6. Buses for the Handicapped (or Dial-a-Ride)

In several major cities a service is available whereby handicapped subscribers can call in several hours beforehand with a request to be taken from A to B at a certain time, with special facilities such as space for a wheel chair if necessary. The short-term problem is to schedule the fleet of specialized mini-buses so as to satisfy a maximum number of requests. One long-term problem is to decide the optimal size of the fleet.

7. Ground Holding of Aircraft

Given several airports, a list of flights, and the capacity of the airports in each period, which is a function of the weather conditions and forecasts, the problem is to decide which planes to delay and by how long, taking into account the numbers of passengers, connecting flights, the expected time until conditions improve, and so on, with the objective of minimizing aircraft costs and passenger inconvenience.

8. Cutting Problems
Whether cutting lengths of paper from rolls, plastic from large rectangular sheets, or patterns to make clothes, the problem is in each case to follow precisely determined cutting rules, satisfy demand, and minimize waste.

Other recent application areas include problems in molecular biology, statistics, and VLSI.
This book tries to provide some of the understanding and tools necessary for tackling such problems.

1.2 WHAT IS AN INTEGER PROGRAM?

Suppose that we have a linear program

$$\max\{cx : Ax \leq b, x \geq 0\}$$

where A is an m by n matrix, c an n-dimensional row vector, b an m-dimensional column vector, and x an n-dimensional column vector of variables or unknowns. Now we add in the restriction that certain variables must take integer values.

If some but not all variables are integer, we have a
(Linear) Mixed Integer Program, written as

$$(MIP) \quad \begin{array}{rl} \max cx + & hy \\ Ax + & Gy \leq b \\ x \geq & 0, y \geq 0 \text{ and integer} \end{array}$$

where A is again m by n, G is m by p, h is a p row-vector, and y is a p column-vector of integer variables.

If all variables are integer, we have a
(Linear) Integer Program, written as

$$(IP) \quad \begin{array}{rl} \max cx \\ Ax \leq & b \\ x \geq & 0 \text{ and integer,} \end{array}$$

and if all variables are restricted to 0–1 values, we have a
0–1 or Binary Integer Program

$$(BIP) \quad \begin{array}{rl} \max cx \\ Ax \leq & b \\ x \in & \{0,1\}^n. \end{array}$$

Another type of problem that we wish to study is a "combinatorial optimization problem." Here typically we are given a finite set $N = \{1,\ldots,n\}$, weights c_j for each $j \in N$, and a set $\mathcal{F}$ of feasible subsets of N. The problem of finding a minimum weight feasible subset is a

Combinatorial Optimization Problem

(COP) $\quad\quad\quad\quad \min_{S \subseteq N}\{\sum_{j \in S} c_j : S \in \mathcal{F}\}.$

In the next section we will see various examples of IPs and $COPs$, and also see that often a COP can be formulated as an IP or BIP.

Given that integer programs look very much like linear programs, it is not surprising that linear programming theory and practice is fundamental in understanding and solving integer programs. However, the first idea that springs to mind, namely "rounding", is often insufficient, as the following example shows.

Example 1.1 Consider the integer program:

$$\max 1.00x_1 + 0.64x_2$$
$$50x_1 + 31x_2 \leq 250$$
$$3x_1 - 2x_2 \geq -4$$
$$x_1, x_2 \geq 0 \text{ and integer.}$$

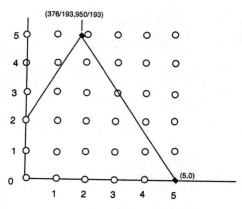

Fig. 1.1 Rounding the LP

As we see from Figure 1.1, the linear programming solution $(376/193, 950/193)$ is a long way from the optimal integer solution $(5, 0)$. ∎

For $BIPs$ the situation is often even worse. The linear programming solution may well be $(0.5, \ldots, 0.5)$, giving no information whatsoever. What is more, it is typically very difficult just to answer the question whether there exists a feasible 0–1 solution.

1.3 FORMULATING IPS AND BIPS

As in linear programming, translating a problem description into a formulation should be done systematically, and a clear distinction should be made between the data of the problem instance, and the variables (or unknowns) used in the model.

(i) Define what appear to be the necessary variables.

(ii) Use these variables to define a set of constraints so that the feasible points correspond to the feasible solutions of the problem.

(iii) Use these variables to define the objective function.

If difficulties arise, define an additional or alternative set of variables and iterate.

Defining variables and constraints may not always be as easy as in linear programming. Especially for $COPs$, we are often interested in choosing a subset $S \subseteq N$. For this we typically make use of the *incidence vector of S*, which is the n-dimensional 0–1 vector x^S such that $x_j^S = 1$ if $j \in S$, and $x_j^S = 0$ otherwise.

Below we formulate four well-known integer programming problems.

The Assignment Problem

There are n people available to carry out n jobs. Each person is assigned to carry out exactly one job. Some individuals are better suited to particular jobs than others, so there is an estimated cost c_{ij} if person i is assigned to job j. The problem is to find a minimum cost assignment.

Definition of the variables.
$x_{ij} = 1$ if person i does job j, and $x_{ij} = 0$ otherwise.

Definition of the constraints.
Each person i does one job:

$$\sum_{j=1}^{n} x_{ij} = 1 \text{ for } i = 1, \ldots, n.$$

Each job j is done by one person:

$$\sum_{i=1}^{n} x_{ij} = 1 \text{ for } j = 1, \ldots, n.$$

The variables are 0–1:

$$x_{ij} \in \{0, 1\} \text{ for } i = 1, \ldots, n, j = 1, \ldots, n.$$

Definition of the objective function.
The cost of the assignment is minimized:
$$\min \sum_{i=1}^{n} \sum_{j=1}^{n} c_{ij} x_{ij}.$$

The 0–1 Knapsack Problem

There is a budget b available for investment in projects during the coming year and n projects are under consideration, where a_j is the outlay for project j, and c_j is its expected return. The goal is to choose a set of projects so that the budget is not exceeded and the expected return is maximized.

Definition of the variables.
$x_j = 1$ if project j is selected, and $x_j = 0$ otherwise.

Definition of the constraints.
The budget cannot be exceeded:
$$\sum_{j=1}^{n} a_j x_j \leq b.$$

The variables are 0–1:
$$x_j \in \{0, 1\} \text{ for } j = 1, \ldots, n.$$

Definition of the objective function.
The expected return is maximized:
$$\max \sum_{j=1}^{n} c_j x_j.$$

The Set Covering Problem

Given a certain number of regions, the problem is to decide where to install a set of emergency service centers. For each possible center the cost of installing a service center, and which regions it can service are known. For instance, if the centers are fire stations, a station can service those regions for which a fire engine is guaranteed to arrive on the scene of a fire within 8 minutes. The goal is to choose a minimum cost set of service centers so that each region is covered.

First we can formulate it as a more abstract COP. Let $M = \{1, \ldots, m\}$ be the set of regions, and $N = \{1, \ldots, n\}$ the set of potential centers. Let $S_j \subseteq M$ be the regions that can be serviced by a center at $j \in N$, and c_j its installation cost. We obtain the problem:
$$\min_{T \subseteq N} \{ \sum_{j \in T} c_j : \cup_{j \in T} S_j = M \}.$$

Now we formulate it as a *BIP*. To facilitate the description, we first construct a 0–1 *incidence matrix* A such that $a_{ij} = 1$ if $i \in S_j$, and $a_{ij} = 0$ otherwise. Note that this is nothing but processing of the data.

Definition of the variables.
$x_j = 1$ if center j is selected, and $x_j = 0$ otherwise.

Definition of the constraints.
At least one center must service region i:

$$\sum_{j=1}^{n} a_{ij} x_j \geq 1 \text{ for } i = 1, \ldots, m.$$

The variables are 0–1:

$$x_j \in \{0, 1\} \text{ for } j = 1, \ldots, n.$$

Definition of the objective function.
The total cost is minimized:

$$\min \sum_{j=1}^{n} c_j x_j.$$

The Traveling Salesman Problem *(TSP)*

This is perhaps the most notorious problem in Operations Research because it is so easy to explain, and so tempting to try and solve. A salesman must visit each of n cities exactly once and then return to his starting point. The time taken to travel from city i to city j is c_{ij}. Find the order in which he should make his tour so as to finish as quickly as possible.

This problem arises in a multitude of forms: a truck driver has a list of clients he must visit on a given day, or a machine must place modules on printed circuit boards, or a stacker crane must pick up and depose crates. Now we formulate it as a *BIP*.

Definition of the variables.
$x_{ij} = 1$ if the salesman goes directly from town i to town j, and $x_{ij} = 0$ otherwise. (x_{ii} is not defined for $i = 1, \ldots, n$.)

Definition of the constraints.
He leaves town i exactly once:

$$\sum_{j: j \neq i} x_{ij} = 1 \text{ for } i = 1, \ldots, n.$$

He arrives at town j exactly once:

$$\sum_{i: i \neq j} x_{ij} = 1 \text{ for } j = 1, \ldots, n.$$

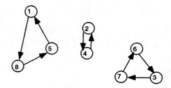

Fig. 1.2 Subtours

So far these are precisely the constraints of the assignment problem. A solution to the assignment problem might give a solution of the form shown in Figure 1.2 (i.e., a set of disconnected subtours). To eliminate these solutions, we need more constraints that guarantee connectivity by imposing that the salesman must pass from one set of cities to another, so-called *cut-set* constraints:

$$\sum_{i \in S} \sum_{j \notin S} x_{ij} \geq 1 \text{ for } S \subset N, S \neq \phi.$$

An alternative is to replace these constraints by *subtour elimination* constraints:

$$\sum_{i \in S} \sum_{j \in S} x_{ij} \leq |S| - 1 \text{ for } S \subset N, 2 \leq |S| \leq n-1.$$

The variables are 0–1:

$$x_{ij} \in \{0,1\} \text{ for } i = 1, \ldots, n, j = 1, \ldots, n, i \neq j.$$

Definition of the objective function.
The total travel time is minimized:

$$\min \sum_{i=1}^{n} \sum_{j=1}^{n} c_{ij} x_{ij}.$$

1.4 THE COMBINATORIAL EXPLOSION

The four problems we have looked at so far are all combinatorial in the sense that the optimal solution is some subset of a finite set. Thus in principle these problems can be solved by enumeration. To see for what size of problem instances this is a feasible approach, we need to count the number of possible solutions.

The Assignment Problem. There is a one-to-one correspondence between assignments and permutations of $\{1, \ldots, n\}$. Thus there are $n!$ solutions to compare.

The Knapsack and Covering Problems. In both cases the number of subsets is 2^n. For the knapsack problem with $b = \sum_{j=1}^{n} a_j/2$, at least half of the subsets are feasible, and thus there are at least 2^{n-1} feasible subsets.

The Traveling Salesman Problem. Starting at city 1, the salesman has $n-1$ choices. For the next choice $n-2$ cities are possible, and so on. Thus there are $(n-1)!$ feasible tours.

In Table 1.1 we show how rapidly certain functions grow. Thus a *TSP* with $n = 101$ has approximately 9.33×10^{157} tours.

n	$\log n$	$n^{0.5}$	n^2	2^n	$n!$
10	3.32	3.16	10^2	1.02×10^3	3.6×10^6
100	6.64	10.00	10^4	1.27×10^{30}	9.33×10^{157}
1000	9.97	31.62	10^6	1.07×10^{301}	4.02×10^{2567}

Table 1.1 Some typical functions

The conclusion to be drawn is that using complete enumeration we can only hope to solve such problems for very small values of n. Therefore we have to devise some more intelligent algorithms, otherwise the reader can throw this book out of the window.

1.5 MIXED INTEGER FORMULATIONS

Modeling Fixed Costs

Suppose we wish to model a typical nonlinear fixed charge cost function:

$$h(x) = f + px \text{ if } 0 < x \leq C \text{ and } h(x) = 0 \text{ if } x = 0$$

with $f > 0$ and $p > 0$ (see Figure 1.3).

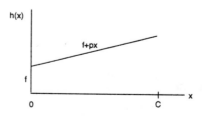

Fig. 1.3 Fixed cost function

Definition of an additional variable.

$$y = 1 \text{ if } x > 0 \text{ and } y = 0 \text{ otherwise.}$$

Definition of the constraints and objective function.
We replace $h(x)$ by $fy + px$, and add the constraints $x \leq Cy, y \in \{0,1\}$.

Note that this is not a completely satisfactory formulation, because although the costs are correct when $x > 0$, it is possible to have the solution $x = 0, y = 1$. However, as the objective is minimization, this will typically not arise in an optimal solution.

Uncapacitated Facility Location *(UFL)*

Given a set of potential depots $N = \{1, \ldots, n\}$ and a set $M = \{1, \ldots, m\}$ of clients, suppose there is a fixed cost f_j associated with the use of depot j, and a transportation cost c_{ij} if all of client i's order is delivered from depot j. The problem is to decide which depots to open, and which depot serves each client so as to minimize the sum of the fixed and transportation costs. Note that this problem is similar to the covering problem, except for the addition of the variable transportation costs.

Definition of the variables.
We introduce a fixed cost or depot opening variable $y_j = 1$ if depot j is used, and $y_j = 0$ otherwise.
x_{ij} is the fraction of the demand of client i satisfied from depot j.

Definition of the constraints.
Satisfaction of the demand of client i:

$$\sum_{j=1}^{n} x_{ij} = 1 \text{ for } i = 1, \ldots, m.$$

To represent the link between the x_{ij} and the y_j variables, we note that $\sum_{i \in M} x_{ij} \leq m$, and use the fixed cost formulation above to obtain:

$$\sum_{i \in M} x_{ij} \leq m y_j \text{ for } j \in N, y_j \in \{0,1\} \text{ for } j \in N.$$

Definition of the objective function.
The objective is $\sum_{j \in N} h_j(x_{1j}, \ldots, x_{mj})$ where $h_j(x_{1j}, \ldots, x_{mj}) = f_j + \sum_{i \in M} c_{ij} x_{ij}$ if $\sum_{i \in M} x_{ij} > 0$, so we obtain

$$\min \sum_{i \in M} \sum_{j \in N} c_{ij} x_{ij} + \sum_{j \in N} f_j y_j.$$

Uncapacitated Lot-Sizing *(ULS)*

The problem is to decide on a production plan for an n-period horizon for a single product. The basic model can be viewed as having data:

- f_t is the fixed cost of producing in period t.
- p_t is the unit production cost in period t.
- h_t is the unit storage cost in period t.
- d_t is the demand in period t.

We use the natural (or obvious) *variables*:

- x_t is the amount produced in period t.
- s_t is the stock at the end of period t.
- $y_t = 1$ if production occurs in t, and $y_t = 0$ otherwise.

To handle the fixed costs, we observe that a priori no upper bound is given on x_t. Thus we either must use a very large value $C = M$, or calculate an upper bound based on the problem data. *fixed cost of producing in period t*
For *constraints* and *objective* we obtain: *1 if we produce in period t*

$$\min \sum_{t=1}^{n} p_t x_t + \sum_{t=1}^{n} h_t s_t + \sum_{t=1}^{n} f_t y_t$$
$$s_{t-1} + x_t = d_t + s_t \text{ for } t = 1, \ldots, n$$
$$x_t \leq My_t \text{ for } t = 1, \ldots, n$$
$$s_0 = 0, s_t, x_t \geq 0, y_t \in \{0, 1\} \text{ for } t = 1, \ldots, n.$$

If we impose that $s_n = 0$, then we can tighten the variable upper bound constraints to $x_t \leq (\sum_{i=t}^{n} d_i) y_t$. Note also that by substituting $s_t = \sum_{i=1}^{t} x_i - \sum_{i=1}^{t} d_i$, the objective function can be rewritten as $\sum_{t=1}^{n} c_t x_t + \sum_{t=1}^{n} f_t y_t - K$ where $c_t = p_t + h_t + \ldots + h_n$ and the constant $K = \sum_{t=1}^{n} h_t (\sum_{i=1}^{t} d_i)$.

Discrete Alternatives or Disjunctions

Suppose $x \in R^n$ satisfies $0 \leq x \leq u$, and either $a^1 x \leq b_1$ or $a^2 x \leq b_2$ (see Figure 1.4). We introduce binary variables y_i for $i = 1, 2$. Then if $M \geq \max\{a^i x - b_i : 0 \leq x \leq u\}$ for $i = 1, 2$, we take as constraints:

$$a^i x - b_i \leq M(1 - y_i) \text{ for } i = 1, 2$$
$$y_1 + y_2 = 1, y_i \in \{0, 1\} \text{ for } i = 1, 2$$
$$0 \leq x \leq u.$$

Now if $y_1 = 1$, x satisfies $a^1 x \leq b_1$ whereas $a^2 x \leq b_2$ is inactive, and conversely if $y_2 = 1$.

Such disjunctions arise naturally in scheduling problems. Suppose that two jobs must be processed on the same machine and cannot be processed

12 FORMULATIONS

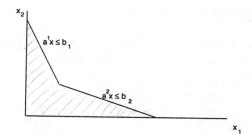

Fig. 1.4 Either/or constraints

simultaneously. If p_i are the processing times, and the variables t_i the start times for $i = 1, 2$, then either job 1 precedes job 2 and so $t_2 \geq t_1 + p_1$, or job 2 comes first and $t_1 \geq t_2 + p_2$.

1.6 ALTERNATIVE FORMULATIONS

In the two previous sections we have formulated a small number of integer programs for which it is not too difficult to verify that the formulations are correct. Here and in the next section we examine alternative formulations and try to understand why some might be better than others. First we make precise what we mean by a formulation.

Definition 1.1 A subset of R^n described by a finite set of linear constraints $P = \{x \in R^n : Ax \leq b\}$ is a *polyhedron*.

Definition 1.2 A polyhedron $P \subseteq R^{n+p}$ is a *formulation* for a set $X \subseteq Z^n \times R^p$ if and only if $X = P \cap (Z^n \times R^p)$. (ie, the polyhedron must contain exactly the same set of integer points)

Note that this definition is restrictive. For example we saw in modeling fixed costs in the previous section that we did not model the set $X = \{(0,0), (x,1)$ for $0 < x \leq C\}$, but the set $X \cup \{(0,1)\}$.

Example 1.2 In Figure 1.5 we show two different formulations for the set:

$$X = \{(1,1), (2,1), (3,1), (1,2), (2,2), (3,2), (2,3)\}.$$ ∎

Note that it would not be easy to write a formulation for the set $X \setminus \{2,2\}$ without introducing many more variables, see Exercise 1.2.

Equivalent Formulations for a 0–1 Knapsack Set

Consider the set of points $X =$

$$\{(0,0,0,0),(1,0,0,0),(0,1,0,0),(0,0,1,0),(0,0,0,1),(0,1,0,1),(0,0,1,1)\}.$$

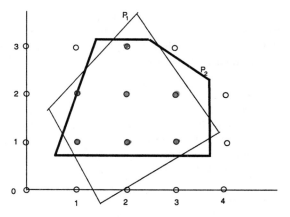

Fig. 1.5 Two different formulations of an IP (Neither is "better" than the other)

Check that the three polyhedra below are formulations for X.

$P_1 = \{x \in R^4 : 0 \leq x \leq 1, 83x_1 + 61x_2 + 49x_3 + 20x_4 \leq 100\}$,

$P_2 = \{x \in R^4 : 0 \leq x \leq 1, 4x_1 + 3x_2 + 2x_3 + 1x_4 \leq 4\}$,

$P_3 = \{x \in R^4 :\ 4x_1\ +3x_2\ +2x_3\ +1x_4\ \leq 4$
$\quad\quad\quad\quad\quad 1x_1\ +1x_2\ +1x_3\ \quad\quad\ \leq 1$
$\quad\quad\quad\quad\quad 1x_1\ \quad\quad\quad\quad\quad\quad +1x_4\ \leq 1$
$\quad\quad\quad\quad\quad 0 \leq x\ \leq 1\}$.

An Equivalent Formulation for **UFL**

For fixed j, consider the constraints:

$$\sum_{i \in M} x_{ij} \leq m y_j,\ y_j \in \{0,1\},\ 0 \leq x_{ij} \leq 1 \text{ for } i \in M.$$

Logically these constraints express the condition: if any $x_{ij} > 0$, then $y_j = 1$, or stated a little differently: for each i, if $x_{ij} > 0$, then $y_j = 1$ (must). This immediately suggests a different set of constraints:

$$0 \leq x_{ij} \leq y_j \text{ for } i \in M, y_j \in \{0,1\}.$$

This leads to the alternative formulation:

$$\min \sum_{i=1}^{m} \sum_{j=1}^{n} c_{ij} x_{ij} + \sum_{j=1}^{n} f_j y_j \quad\quad (1.1)$$
$$\sum_{j=1}^{n} x_{ij} = 1 \text{ for } i \in M \quad\quad (1.2)$$
$$x_{ij} \leq y_j \text{ for } i \in M, j \in N \quad\quad (1.3)$$
$$x_{ij} \geq 0 \text{ for } i \in M, j \in N, y_j \in \{0,1\} \text{ for } j \in N. \quad\quad (1.4)$$

In the two examples we have just examined, the variables have been the same in each formulation, and we have essentially modified or added constraints. Another possible approach is to add or choose different variables, in which case we talk of *extended formulations*.

An Extended Formulation for *ULS*

What variables, other than production and stock levels, might be useful in describing an optimal solution of the lot-sizing problem? Suppose we would like to know when the items being produced now will actually be used to satisfy demand. Following the same steps as before, this leads to:

Definition of the variables.
w_{it} is the amount produced in period i to satisfy demand in period t
$y_t = 1$ if production occurs in period t (as above), and $y_t = 0$ otherwise.

Definition of the constraints.
Satisfaction of demand in period t:

$$\sum_{i=1}^{t} w_{it} = d_t \text{ for all } t.$$

Variable upper-bound constraints:

$$w_{it} \leq d_t y_i \text{ for all } i, t (i \leq t).$$

The variables are mixed:

$$w_{it} \geq 0 \text{ for all } i, t, i \leq t, y_t \in \{0, 1\} \text{ for all } t.$$

Definition of the objective function.

$$\min \sum_{i=1}^{n} \sum_{t=i}^{n} c_i w_{it} + \sum_{t=1}^{n} f_t y_t.$$

Note that here we are again using modified costs with c_t the production cost in period t and zero storage costs. It is optional whether we choose to define the old variables in terms of the new by adding defining equalities:

$$x_i = \sum_{t=i}^{n} w_{it}.$$

1.7 GOOD AND IDEAL FORMULATIONS

In Figure 1.5 we show two different formulations of the same problem. We have also seen two possible formulations for the uncapacitated facility location problem. Geometrically we can see that there must be an infinite number of formulations, so how can we choose between them?

The geometry again helps us to find an answer. Look at Figure 1.6 in which we have repeated the two formulations shown in Figure 1.5, and added a third one P_3. Formulation P_3 is *ideal*, because now if we solve a linear program over P_3, the optimal solution is at an extreme point. In this ideal case each extreme point is integer and so the IP is solved.

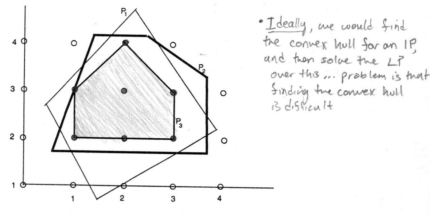

Fig. 1.6 The ideal formulation

- Ideally, we would find the convex hull for an IP, and then solve the LP over this ... problem is that finding the convex hull is difficult

We can now formalize this idea.

Definition 1.3 Given a set $X \subseteq R^n$, the *convex hull of* X, denoted conv(X), is defined as: $conv(X) = \{x : x = \sum_{i=1}^{t} \lambda_i x^i, \sum_{i=1}^{t} \lambda_i = 1, \lambda_i \geq 0$ for $i = 1, \ldots, t$ over all finite subsets $\{x^1, \ldots, x^t\}$ of $X\}$.

→ Or just use the extreme points

Proposition 1.1 $conv(X)$ *is a polyhedron*.

Proposition 1.2 *The extreme points of* $conv(X)$ *all lie in* X.

Because of these two results, we can replace the IP: $\{\max cx : x \in X\}$ by the equivalent linear program: $\{\max cx : x \in conv(X)\}$. This ideal reduction to a linear program also holds for unbounded integer sets $X = \{x : Ax \leq b, x \geq 0$ and integer$\}$, and mixed integer sets $X = \{(x,y) : Ax + Gy \leq b, x \geq 0, y \geq 0$ and integer$\}$ with A, G, b rational. However, whether X is bounded or not, this is in general only a theoretical solution, because in most cases there is such an enormous (exponential) number of inequalities needed to describe conv(X), and there is no simple characterization for them.

So we might rather ask the question: Given two formulations P_1 and P_2 for X, when can we say that one is *better* than the other? Because the ideal solution conv(X) has the property that $X \subseteq conv(X) \subseteq P$ for all formulations P, this suggests the following definition.

Definition 1.4 Given a set $X \subseteq R^n$, and two formulations P_1 and P_2 for X, P_1 is a *better formulation* than P_2 if $P_1 \subset P_2$.

In the next chapter we will see that this turns out to be a useful definition, and a very important idea for the development of effective formulations. For the moment we examine some of our formulations in the light of these definitions.

Equivalent Formulations for a Knapsack Set

Looking again at the set $X =$
$\{(0,0,0,0),(1,0,0,0),(0,1,0,0),(0,0,1,0),(0,0,0,1),(0,1,0,1),(0,0,1,1)\}$
and the three formulations,

$$P_1 = \{x \in R^4 : 0 \leq x \leq 1, 83x_1 + 61x_2 + 49x_3 + 20x_4 \leq 100\},$$

$$P_2 = \{x \in R^4 : 0 \leq x \leq 1, 4x_1 + 3x_2 + 2x_3 + 1x_4 \leq 4\},$$

$$P_3 = \{x \in R^4 : \begin{array}{llllr} 4x_1 & +3x_2 & +2x_3 & +1x_4 & \leq 4 \\ 1x_1 & +1x_2 & +1x_3 & & \leq 1 \\ 1x_1 & & & +1x_4 & \leq 1 \\ & & & 0 \leq x & \leq 1\}, \end{array}$$

it is easily seen that $P_3 \subset P_2 \subset P_1$, and it can be checked that $P_3 = \text{conv}(X)$, and P_3 is thus an ideal formulation.

Formulations for *UFL*

Let P_1 be the formulation given in Section 1.4 with a single constraint

$$\sum_{i \in M} x_{ij} \leq m y_j$$

for each j, and P_2 be the formulation given in Section 1.5 with the m constraints

$$x_{ij} \leq y_j \text{ for } i \in M$$

for each j. Note that if a point (x, y) satisfies the constraints $x_{ij} \leq y_j$ for $i \in M$, then summing over $i \in M$ shows that it also satisfies the constraints $\sum_i x_{ij} \leq m y_j$. Thus $P_2 \subseteq P_1$. To show that $P_2 \subset P_1$, we need to find a point in P_1 that is not in P_2. Suppose for simplicity that n divides m, that is, $m = kn$ with $k \geq 2$ and integer. Then a point in which each depot serves k clients, $x_{ij} = 1$ for $i = k(j-1)+1, \ldots, k(j-1)+k, j = 1, \ldots, n$, $x_{ij} = 0$ otherwise, and $y_j = k/m$ for $j = 1, \ldots, n$ lies in $P_1 \setminus P_2$.

The two formulations that we have seen for the lot-sizing problem *ULS* have different variables, so it is not immediately obvious how they can be compared. We now examine this question briefly.

GOOD AND IDEAL FORMULATIONS

Formally, assuming for simplicity that all the variables are integer, we have a first formulation
$$\min\{cx : x \in P \cap Z^n\}$$
with $P \subseteq R^n$, and a second extended formulation
$$\min\{cx : (x,w) \in Q \cap (Z^n \times R^p)\}$$
with $Q \subseteq R^n \times R^p$. ← we want to project this onto the common space R^n

Definition 1.5 Given a polyhedron $Q \subseteq (R^n \times R^p)$, the *projection of Q* onto the subspace R^n, denoted $proj_x Q$, is defined as:
$$proj_x Q = \{x \in R^n : (x,w) \in Q \text{ for some } w \in R^p\}.$$

Thus $proj_x Q \subseteq R^n$ is the formulation obtained from Q in the space R^n of the original x variables, and it allows us to compare Q with other formulations $P \subseteq R^n$.

Comparing Formulations for ULS ← with elements $(\vec{x}, \vec{s}, \vec{y})$

We can now compare the formulation P_1:
$$\begin{aligned} s_{t-1} + x_t &= d_t + s_t \text{ for } t = 1, \ldots, n \\ x_t &\leq M y_t \text{ for } t = 1, \ldots, n \\ s_0 = 0, s_t, x_t &\geq 0, 0 \leq y_t \leq 1 \text{ for all } t, \end{aligned}$$

and $P_2 = proj_{x,s,y} Q_2$, where Q_2 takes the form:

$$\begin{aligned} \sum_{i=1}^{t} w_{it} &= d_t \text{ for all } t \\ w_{it} &\leq d_t y_i \text{ for all } i, t, i \leq t \\ x_i &= \sum_{t=i}^{n} w_{it} \text{ for all } i \\ w_{it} &\geq 0 \text{ for all } i, t \text{ with } i \leq t \\ 0 &\leq y_t \leq 1 \text{ for all } t. \end{aligned}$$

It can be shown that P_2 describes the convex hull of solutions to the problem, and is thus an ideal formulation. It is easily seen that $P_2 \subset P_1$. For instance the point $x_t = d_t, y_t = d_t/M$ for all t is an extreme point of P_1 that is not in P_2.

↑ No storage ever!   ↑ production in every period

1.8 NOTES

Introduction

Much of the material in this book except for the last three or four chapters is treated in greater detail and at a more advanced level in [NemWol88]. An even more advanced theoretical treatment of integer programming is given in [Sch86]. Other recent books include [ParRar88], [SalMat89] and [Sie96]. For the related topic of combinatorial optimization, an undergraduate text is [PapSti82], and the recent graduate-level book [CooCunetal97] is recommended.

An important source of references and surveys are the annotated bibliographies of [OheLenRin85] and [DelAMafMar97]. The journals publishing articles on integer programming cover a wide spectrum from theory to applications. The more theoretical include *Mathematical Programming, Mathematics of Operations Research, the SIAM Journal on Discrete Mathematics, the SIAM Journal on Optimization* and *Discrete Applied Mathematics*, while *Operations Research, the European Journal of Operations Research, Networks, Management Science*, and so on, contain more integer programming applications. For a survey of integer programming, see [NemWol89].

For those unfamiliar with linear programming , [Dan63] is the ultimate classic, and the book [Chv83] is exceptional. Recently many books treating interior point as well as simplex methods have appeared, including [RooTerVia97] and [Vdbei96]. For graph theory the books [Ber73] and [BonMur76] are classics. For network flows [ForFul62] and [Law76] remain remarkably stimulating, while [AhuMagOrl93] is very comprehensive and up-to-date.

In the notes at the end of each chapter, we will cite only selected references: a few of the important original sources, and also some recent surveys and articles that can be used for further reading.

Notes for the Chapter

The importance of formulations in integer programming has only become fully apparent in the last twenty years. The book [Wil78] is partly devoted to this topic, but is now a little dated. Certain special classes of problems that are often tackled as integer programs have had books dedicated to them: knapsack problems [MarTot90], traveling salesman problems [Lawetal85], location problems [MirFra90], network flows [AhuMagOrl93], network models [Balletal95a] and network routing [Balletal95b], and production and inventory [GraRinZip93]. For scheduling problems the recent survey [QueSch94] contains a variety of integer programming formulations.

1.9 EXERCISES

1. Suppose that you are interested in choosing a set of investments $\{1, \ldots, 7\}$ using 0–1 variables. Model the following constraints:

Let $x_i = 1$ is you invest in asset i

(i) You cannot invest in all of them. $\sum_{i=1}^{7} x_i \leq 6$
(ii) You must choose at least one of them. $\sum_{i=1}^{7} x_i \geq 1$
(iii) Investment 1 cannot be chosen if investment 3 is chosen. $x_1 \leq 1 - x_3$
(iv) Investment 4 can be chosen only if investment 2 is also chosen. $x_4 \leq x_2$
(v) You must choose either both investments 1 and 5 or neither. $x_1 = x_5$
(vi) You must choose either at least one of the investments 1,2,3, or at least two investments from 2,4,5,6.

2. Formulate the following as mixed integer programs:

(i) $u = \min\{x_1, x_2\}$, assuming that $0 \leq x_j \leq C$ for $j = 1, 2$
(ii) $v = |x_1 - x_2|$ with $0 \leq x_j \leq C$, for $j = 1, 2$
(iii) the set $X \setminus \{x^*\}$ where $X = \{x \in Z^n : Ax \leq b\}$ and $x^* \in X$.

3. Modeling disjunctions.

(i) Extend the formulation of discrete alternatives of Section 1.4 to the union of two polyhedra $P_k = \{x \in R^n : A^k x \leq b^k, 0 \leq x \leq u\}$ for $k = 1, 2$ where $\max_k \max_i \{a_i^k x - b_i^k : 0 \leq x \leq u\} \leq M$.
(ii) Show that an extended formulation for $P_1 \cup P_2$ is

$$x = z^1 + z^2$$
$$A^k z^k \leq b^k y^k \text{ for } k = 1, 2$$
$$0 \leq z^k \leq u y^k \text{ for } k = 1, 2$$
$$y^1 + y^2 = 1$$
$$z^k \in R^n, y^k \in B^1 \text{ for } k = 1, 2.$$

4. Show that

$$X = \{x \in B^4 : 97x_1 + 32x_2 + 25x_3 + 20x_4 \leq 139\}$$
$$= \{x \in B^4 : 2x_1 + x_2 + x_3 + x_4 \leq 3\}$$
$$= \{x \in B^4 : x_1 + x_2 + x_3 \leq 2$$
$$x_1 + x_2 + x_4 \leq 2$$
$$x_1 + x_3 + x_4 \leq 2\}.$$

5. John Dupont is attending a summer school where he must take four courses per day. Each course lasts an hour, but because of the large number of students, each course is repeated several times per day by different teachers. Section i of course k denoted (i, k) meets at the hour t_{ik}, where courses start on the hour between 10 a.m. and 7 p.m. John's preferences for when he takes courses are influenced by the reputation of the teacher, and also the time of

day. Let p_{ik} be his preference for section (i,k). Unfortunately, due to conflicts, John cannot always choose the sections he prefers.

(i) Formulate an integer program to choose a feasible course schedule that maximizes the sum of John's preferences.
(ii) Modify the formulation, so that John never has more than two consecutive hours of classes without a break.
(iii) Modify the formulation, so that John chooses a schedule in which he starts his day as late as possible.

6. Prove that the set of feasible solutions to the formulation of the traveling salesman problem in Section 1.2 is precisely the set of incidence vectors of tours.

7. The QED Company must draw up a production program for the next nine weeks. Jobs last several weeks and once started must be carried out without interruption. During each week a certain number of skilled workers are required to work full-time on the job. Thus if job i lasts p_i weeks, $l_{i,u}$ workers are required in week u for $u = 1, \ldots, p_i$. The total number of workers available in week t is L_t. Typical job data $(i, p_i, l_{i1}, \cdots, l_{ip_i})$ is shown below.

Job	Length	Week1	Week2	Week3	Week4
1	3	2	3	1	-
2	2	4	5	-	-
3	4	2	4	1	5
4	4	3	4	2	2
5	3	9	2	3	-

(i) Formulate the problem of finding a feasible schedule as an IP.
(ii) Formulate when the objective is to minimize the maximum number of workers used during any of the nine weeks.
(iii) Job 1 must start at least two weeks before job 3. Formulate.
(iv) Job 4 must start not later than one week after job 5. Formulate.
(v) Jobs 1 and 2 both need the same machine, and cannot be carried out simultaneously. Formulate.

8. Show that the uncapacitated facility location problem of Section 1.4 with a set $N = \{1, \ldots, n\}$ of depots can be written as the COP

$$\min_{S \subseteq N} \{c(S) + \sum_{j \in S} f_j\}$$

where $c(S) = \sum_{i=1}^{m} \min_{j \in S} c_{ij}$.

9. Show that the covering problem of Section 1.2 can be written as the COP

$$\min_{S \subseteq N} \{\sum_{j \in S} f_j : v(S) = v(N)\}$$

where $v(S) = \sum_{i=1}^{m} \min\{\sum_{j \in S} a_{ij}, 1\}$.

10. A set of n jobs must be carried out on a single machine that can do only one job at a time. Each job j takes p_j hours to complete. Given job weights w_j for $j = 1, \ldots, n$, in what order should the jobs be carried out so as to minimize the weighted sum of their start times? Formulate this scheduling problem as a mixed integer program.

11. Using a mixed integer programming system, solve an instance of the uncapacitated facility location problem, where f_j is the cost of opening depot j, and c_{ij} is the cost of satisfying all client i's demand from depot j, with $f = (4, 3, 4, 4, 7)$ and

$$(c_{ij}) = \begin{pmatrix} 12 & 13 & 6 & 0 & 1 \\ 8 & 4 & 9 & 1 & 2 \\ 2 & 6 & 6 & 0 & 1 \\ 3 & 5 & 2 & 1 & 8 \\ 8 & 0 & 5 & 10 & 8 \\ 2 & 0 & 3 & 4 & 1 \end{pmatrix}.$$

12. The symmetric traveling salesman problem is a TSP in which $c_{ij} = c_{ji}$ for all $(i, j) \in A$. Consider an instance on the graph shown in Figure 1.7, where the missing edges have a very high cost. Solve with a mixed integer programming system.

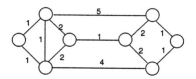

Fig. 1.7 TSP instance

13. Formulate and solve an instance of the lot-sizing problem over 6 periods, with demands $(6, 7, 4, 6, 3, 8)$, unit production costs $(3, 4, 3, 4, 4, 5)$, unit storage costs $(1, 1, 1, 1, 1, 1)$, set-up costs $(12, 15, 30, 23, 19, 45)$, and a maximum production capacity of 10 items per period.

14. Formulate and solve the problem of placing N queens on an N by N chessboard such that no two queens share any row, column, or diagonal.

15. (Projection). Let $Q = \{(x, y) \in R^n_+ \times R^p_+ : Ax + Gy \leq b\}$. Use Farkas' Lemma to show that

$$proj_y(Q) = \{y \in R^p_+ : v^t(b - Gy) \geq 0 \text{ for } t = 1, \ldots, T\}$$

where $\{v^t\}_{t=1}^T$ are the extreme rays of $V = \{v \in R^m_+ : vA \geq 0\}$.

Show that if $proj_y(Q) \neq \emptyset$,

$$\max\{cx + hy : (x, y) \in Q\} = \max_y \{\min_{s=1,\cdots,S} u^s(b - Gy) + hy : y \in proj_y(Q)\}$$

where $\{u^s\}_{s=1}^S$ are the extreme points of $U = \{u \in R^m_+ : uA \geq c\}$.

2
Optimality, Relaxation, and Bounds

2.1 OPTIMALITY AND RELAXATION

Given an IP or COP

$$z = \max\{c(x) : x \in X \subseteq Z^n\},$$

how is it possible to prove that a given point x^* is optimal? Put differently, we are looking for some optimality conditions that will provide stopping criteria in an algorithm for IP.

The "naive" but nonetheless important reply is that we need to find a lower bound $\underline{z} \leq z$ and an upper bound $\overline{z} \geq z$ such that $\underline{z} = \overline{z} = z$. Practically, this means that any algorithm will find a decreasing sequence

$$\overline{z_1} > \overline{z_2} > \ldots > \overline{z_s} \geq z$$

of upper bounds, and an increasing sequence

$$\underline{z_1} < \underline{z_2} < \ldots < \underline{z_t} \leq z,$$

of lower bounds, and stop when

$$\overline{z_s} - \underline{z_t} \leq \epsilon,$$

where ϵ is some suitably chosen small nonnegative value (see Figure 2.1). Thus we need to find ways of deriving such upper and lower bounds.

24 OPTIMALITY, RELAXATION, AND BOUNDS

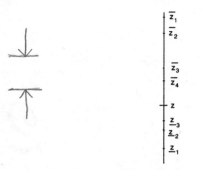

Fig. 2.1 Bounds for IP *(for a maximization problem)*

Primal Bounds

Every feasible solution $x^* \in X$ provides a lower bound $\underline{z} = c(x^*) \leq z$. This is essentially the only way we know to obtain lower bounds. For some IP problems, finding feasible solutions is easy, and the real question is how to find good solutions. For instance in the traveling saleman problem, if the salesman is allowed to travel between any pair of cities, any permutation of the cities leads to a feasible tour, and it suffices to evaluate the length of the tour to have a primal bound on z. Some simple ways to find feasible solutions and then improve them are discussed later in this chapter. For other IPs, finding feasible solutions may be very difficult (as difficult as the IP itself). This topic is raised again when we discuss complexity in Chapter 6, and heuristics to find primal bounds are treated in more detail in Chapter 12.

Dual Bounds

Finding upper bounds for a maximization problem (or lower bounds for a minimization problem) presents a different challenge. These are called *dual bounds* in contrast to the primal bounds for reasons that should become obvious in Section 2.5. The most important approach is by "relaxation," the idea being to replace a "difficult" max(min) IP by a simpler optimization problem whose optimal value is at least as large (small) as z.

For the "relaxed" problem to have this property, there are two obvious possibilities:

(i) Enlarge the set of feasible solutions so that one optimizes over a larger set, or
(ii) Replace the max(min) objective function by a function that has the same or a larger (smaller) value everywhere.

Definition 2.1 A problem (RP) $z^R = \max\{f(x) : x \in T \subseteq R^n\}$ is a *relaxation* of (IP) $z = \max\{c(x) : x \in X \subseteq R^n\}$ if :

(i) $X \subseteq T$, and
(ii) $f(x) \geq c(x)$ for all $x \in X$.

Proposition 2.1 *If RP is a relaxation of IP, $z^R \geq z$.*

Proof. If x^* is an optimal solution of IP, $x^* \in X \subseteq T$ and $z = c(x^*) \leq f(x^*)$. As $x^* \in T$, $f(x^*)$ is a lower bound on z^R, and so $z \leq f(x^*) \leq z^R$. ∎

The question then arises of how to construct interesting relaxations. One of the most useful and natural ones is the linear programming relaxation.

2.2 LINEAR PROGRAMMING RELAXATIONS

Definition 2.2 For the integer program $\max\{cx : x \in P \cap Z^n\}$ with formulation $P = \{x \in R_+^n : Ax \leq b\}$, the *linear programming relaxation* is the linear program $z^{LP} = \max\{cx : x \in P\}$.

As $P \cap Z^n \subseteq P$ and the objective function is unchanged, this is clearly a relaxation.

Example 2.1 Consider the integer program

$$\begin{aligned} z = \max \quad & 4x_1 - x_2 \\ & 7x_1 - 2x_2 \leq 14 \\ & x_2 \leq 3 \\ & 2x_1 - 2x_2 \leq 3 \\ & x \in Z_+^2. \end{aligned}$$

To obtain a primal (lower) bound, observe that $(2,1)$ is a feasible solution, so we have the lower bound $z \geq 7$. To obtain a dual (upper) bound, consider the linear programming relaxation. The optimal solution is $x^* = (\frac{20}{7}, 3)$ with value $z^{LP} = \frac{59}{7}$. Thus we obtain an upper bound $z \leq \frac{59}{7}$. Observing that the optimal value must be integer, we can round down to the nearest integer and so obtain $z \leq 8$. ∎

Note that the definition of better formulations is intimately related to that of linear programming relaxations. In particular better formulations give tighter (dual) bounds.

Proposition 2.2 *Suppose P_1, P_2 are two formulations for the integer program $\max\{cx : x \in X \subseteq Z^n\}$ with P_1 a better formulation than P_2 i.e. $P_1 \subset P_2$. If $z_i^{LP} = \max\{cx : x \in P^i\}$ for $i = 1,2$ are the values of the associated linear programming relaxations, then $z_1^{LP} \leq z_2^{LP}$ for all c.*

Relaxations do not just give dual bounds. They sometimes allow us to prove optimality.

26 OPTIMALITY, RELAXATION, AND BOUNDS

Proposition 2.3 (i) *If a relaxation RP is infeasible, the original problem IP is infeasible.*
(ii) *Let x^* be an optimal solution of RP. If $x^* \in X$ and $f(x^*) = c(x^*)$, then x^* is an optimal solution of IP.*

Proof. (i) As RP is infeasible, $T = \phi$ and thus $X = \phi$.
(ii) As $x^* \in X, z \geq c(x^*) = f(x^*) = z^R$. As $z \leq z^R$, we obtain $c(x^*) = z = z^R$. ∎

Example 2.2 The linear programming relaxation of the integer program:
$$\max 7x_1 + 4x_2 + 5x_3 + 2x_4$$
$$3x_1 + 3x_2 + 4x_3 + 2x_4 \leq 6$$
$$x \in B^4$$
has optimal solution $x^* = (1, 1, 0, 0)$. As x^* is integral, it solves the integer program. ∎

2.3 COMBINATORIAL RELAXATIONS

Whenever the relaxed problem is a combinatorial optimization problem, we speak of a *combinatorial relaxation*. In many cases, such as (i)–(iii) below, the relaxation is an easy problem that can be solved rapidly. Some of the problems arising in this way are studied in Chapters 3 and 4. Here we illustrate with four examples.

(i) **The Traveling Salesman Problem.** We saw in formulating the traveling salesman problem with digraph $D = (V, A)$ and arc weights c_{ij} for $(i, j) \in A$ that the (Hamiltonian or salesman) tours are precisely the assignments (or permutations) containing no subtours. Thus
$$z^{TSP} = \min_{T \subseteq A} \{ \sum_{(i,j) \in T} c_{ij} : T \text{ forms a tour} \} \geq$$
$$z^{ASS} = \min_{T \subseteq A} \{ \sum_{(i,j) \in T} c_{ij} : T \text{ forms a assignment} \}.$$

(ii) A closely related problem is the **Symmetric Traveling Salesman Problem (STSP)** specified by a graph $G = (V, E)$ and edge weights c_e for $e \in E$. The problem is to find an undirected tour of minimum weight. An interesting relaxation of this problem is obtained by observing that

a) Every tour consists of two edges adjacent to node 1, and a path through nodes $\{2, \ldots, n\}$.

b) A path is a special case of a tree.

Definition 2.3 A *1-tree* is a subgraph consisting of two edges adjacent to node 1, plus the edges of a tree on nodes $\{2, \ldots, n\}$.

Clearly every tour is a 1-tree, and thus {tours} ⊆ {1-trees}

$$z^{STSP} = \min_{T \subseteq E} \{\sum_{e \in T} c_e : T \text{ forms a tour}\} \geq$$

$$z^{1-tree} = \min_{T \subseteq E} \{\sum_{e \in T} c_e : T \text{ forms a 1-tree}\}.$$

(iii) **The Quadratic 0–1 Problem** is the problem:

$$\max\{\sum_{i,j: 1 \leq i < j \leq n} q_{ij} x_i x_j - \sum_{j=1}^{n} p_j x_j, x \neq \mathbf{0}, x \in B^n\}.$$

Replacing all terms $q_{ij} x_i x_j$ with $q_{ij} < 0$ by 0 gives a relaxation

$$z^R = \max\{\sum_{i,j: 1 \leq i < j \leq n} \max\{q_{ij}, 0\} x_i x_j - \sum_{j=1}^{n} p_j x_j, x \neq \mathbf{0}, x \in B^n\}.$$

In Chapter 9, it will be shown how this relaxation can be solved as a series of maximum flow problems.

(iv). **The Knapsack Problem.** A relaxation of the set $X = \{x \in Z_+^n : \sum_{j=1}^{n} a_j x_j \leq b\}$ is the set

$$X' = \{x \in Z_+^n : \sum_{j=1}^{n} \lfloor a_j \rfloor x_j \leq \lfloor b \rfloor\}$$

where $\lfloor a \rfloor$ is the largest integer less than or equal to a.

2.4 LAGRANGIAN RELAXATION

Suppose we are given an integer program (IP) in the form $z = \max\{cx : Ax \leq b, x \in X \subseteq Z^n\}$. If the problem is too difficult to solve directly, one possiblity is just to drop the constraints $Ax \leq b$. Clearly the resulting problem: $z' = \max\{cx : x \in X\}$ is a relaxation of IP. In the asymmetric traveling salesman problem above, the assignment problem is obtained by dropping the subtour constraints. An important extension of this idea, dealt with in much greater detail in Chapter 10, is not just to drop complicating constraints, but then to add them into the objective function with Lagrange multipliers (dual variables).

Proposition 2.4 Let $z(u) = \max\{cx + u(b - Ax) : x \in X\}$. Then $z(u) \geq z$ for all $u \geq 0$.

(Ax ≤ b ⟹ 0 ≤ b − Ax)

- Clearly this a relaxation, since constraints have been removed, and the prop shows that $z(u) \geq z$

Proof. Let x^* be optimal for IP. As x^* is feasible in IP, $x^* \in X$. Again by feasibility $Ax^* \leq b$, and thus as $u \geq 0$, $cx^* \leq cx^* + u(b - Ax^*) \leq z(u)$ where the last inequality is by definition of $z(u)$. ∎

2.5 DUALITY

For linear programs duality provides a standard way to obtain upper bounds. It is therefore natural to ask whether it is possible to find duals for integer programs. The important property of a dual is that the value of any feasible solution provides an upper bound on the objective value z. This suggests the following definition. *(for a max problem)*

Definition 2.4 The two problems

$$(IP) \qquad z = \max\{c(x) : x \in X\}$$

$$(D) \qquad w = \min\{\omega(u) : u \in U\}$$

form a *(weak)-dual pair* if $c(x) \leq \omega(u)$ for all $x \in X$ and all $u \in U$. When $z = w$, they form a *strong-dual pair*.

The advantage of a dual problem as opposed to a relaxation is that any dual feasible solution provides an upper bound on z, whereas a relaxation of IP must be solved to optimality to provide such a bound. Do such dual problems exist?

Not surprisingly, a linear programming relaxation immediately leads to a weak dual.

Proposition 2.5 *The integer program $z = \max\{cx : Ax \leq b, x \in Z_+^n\}$ and the linear program $w^{LP} = \min\{ub : uA \geq c, u \in R_+^m\}$ form a weak dual pair.*

By analogy with Proposition 2.3, dual problems sometimes allow us to prove optimality.

Proposition 2.6 *Suppose that IP and D are a weak-dual pair.*
(i) If D is unbounded, IP is infeasible.
(ii) If $x^* \in X$ and $u^* \in U$ satisfy $c(x^*) = \omega(u^*)$, then x^* is optimal for IP and u^* is optimal for D.

We now present another example of a dual pair of problems.

A Matching Dual

Given a graph $G = (V, E)$, a *matching* $M \subseteq E$ is a set of disjoint edges. A *covering by nodes* is a set $R \subseteq V$ of nodes such that every edge has at least one endpoint in R.

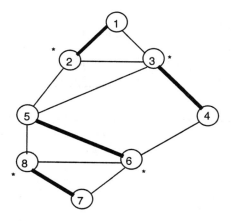

Fig. 2.2 Matching and cover by nodes

A graph on eight nodes is shown in Figure 2.2. The edges $(1,2), (3,4), (5,6)$, and $(7,8)$ form a matching and the nodes $\{2,3,6,8\}$ a cover by nodes.

Proposition 2.7 *The problem of finding a <u>maximum cardinality matching</u>:*

$$\max_{M \subseteq E}\{|M| : M \text{ is a matching }\}$$

and the problem of finding a <u>minimum cardinality covering by nodes</u>:

$$\min_{R \subseteq V}\{|R| : R \text{ is a covering by nodes }\}$$

<u>form a weak-dual pair</u>.

Proof. If M is a matching with $M = \{(i_1, j_1), \ldots, (i_k, j_k)\}$, then the $2k$ nodes $\{i_1, j_1, \ldots, i_k, j_k\}$ are distinct, and any covering by nodes R must contain at least one node from each pair $\{i_s, j_s\}$ for $s = 1, \ldots, k$. Therefore $|R| \geq k = |M|$. ∎

We can also establish this result using linear programming duality.

Definition 2.5 The *node-edge incidence matrix* of a graph $G = (V, E)$ is an $n = |V|$ by $m = |E|$ 0–1 matrix A with $a_{j,e} = 1$ if node j is an endpoint of edge e, and $a_{j,e} = 0$ otherwise.

The maximum cardinality matching problem can now be formulated as the integer program:

$$z = \max\{\mathbf{1}x : Ax \leq 1, x \in Z_+^m\}$$

and the minimum cardinality covering by nodes problem as:

$$w = \min\{\mathbf{1}y : yA \geq 1, y \in Z_+^n\}.$$

Let z^{LP} and w^{LP} be the values of their corresponding linear programming relaxations. Then $z \leq z^{LP} = w^{LP} \leq w$, and the duality is again established.

Example 2.3 It is easily seen that there is not a strong duality between the two problems. Consider the graph shown in Figure 2.3.

Fig. 2.3 Weak duality for matching

First observe that $z = 1$ and $w = 2$. What is more, $x_{e_1} = x_{e_2} = x_{e_3} = 1/2$ is feasible for the first LP relaxation, and $y_1 = y_2 = y_3 = 1/2$ is feasible for the second LP relaxation, so $z^{LP} = w^{LP} = 3/2$. ∎

Later we will see that strong duality holds for this pair of problems when the graph G is bipartite.

2.6 PRIMAL BOUNDS: GREEDY AND LOCAL SEARCH

Now we briefly consider how to obtain primal bounds. The idea of a *greedy heuristic* is to construct a solution from scratch (the empty set), choosing at each step the item bringing the "best" immediate reward. We present two examples.

Example 2.4 The 0–1 Knapsack Problem. Consider the instance:

$$\max 12x_1 + 8x_2 + 17x_3 + 11x_4 + 6x_5 + 2x_6 + 2x_7$$
$$4x_1 + 3x_2 + 7x_3 + 5x_4 + 3x_5 + 2x_6 + 3x_7 \leq 9$$
$$x \in B^7.$$

Noting that the variables are ordered so that $\frac{c_j}{a_j} \geq \frac{c_{j+1}}{a_{j+1}}$ for $j = 1, \ldots, n-1$, a greedy solution is:
(i) As $\frac{c_1}{a_1}$ is maximum and there is enough space (9 units) available, fix $x_1 = 1$.
(ii) In the remaining problem, as $\frac{c_2}{a_2}$ is maximum and there is enough space (5=9-4) units available, fix $x_2 = 1$.
(iii) As each item 3,4,5 in that order requires more space than the $2 = 5 - 3$ units available, set $x_3 = x_4 = x_5 = 0$.
(iv) As $\frac{c_6}{a_6}$ is maximum in the remaining problem, and there is enough space (2 units) available, fix $x_6 = 1$.
(v) Set $x_7 = 0$ as no further space is available.

Therefore the greedy solution is $x^G = (1, 1, 0, 0, 0, 1, 0)$ with value $z^G = cx^G = 32$. ∎

Example 2.5 The Symmetric Traveling Salesman Problem. Consider an instance with distance matrix:

$$(c_e) = \begin{matrix} & 1 & 2 & 3 & 4 & 5 & 6 \\ 1 & - & 9 & 2 & 8 & 12 & 11 \\ 2 & & - & 7 & 19 & 10 & 32 \\ 3 & & & - & 29 & 18 & 6 \\ 4 & & & & - & 24 & 3 \\ 5 & & & & & - & 19 \\ 6 & & & & & & - \end{matrix}$$

Greedy examines the edges in order of nondecreasing length.

The cheapest edge is $e_1 = (1,3)$ with $c_{e_1} = 2$. Select the edge by setting $x_{e_1} = 1$.

The next cheapest edge remaining is $e_2 = (4,6)$ with $c_{e_2} = 3$. As edges e_1 and e_2 can appear together in a tour, set $x_{e_2} = 1$.

Set $x_{e_3} = 1$ where $e_3 = (3,6)$ and $c_{e_3} = 6$.

Set $x_{e_4} = 0$ where $e_4 = (2,3)$ as node 3 already has degree 2, and all three edges $(1,3),(3,6),(2,3)$ cannot be in a tour.

Set $x_{e_5} = 0$ where $e_5 = (1,4)$ as edge $(1,4)$ forms a subtour with the edges already chosen.

Set $x_{e_6} = 1$ where $e_6 = (1,2)$ and $c_{e_6} = 9$. (This will be our return to start edge)

Continue as above, choosing edges $(2,5)$ with length 10 and $(4,5)$ with length 24 to complete the tour.

The greedy tour is $(1,3,6,4,5,2,1)$ with length $\sum_e c_e x_e = 54$. ∎

Once an initial feasible solution, called the *incumbent*, has been found, it is natural to try and improve the solution. The idea of a *local search heuristic* is to define a neighborhood of solutions close to the incumbent. Then the best solution in the neighborhood is found. If it is better than the incumbent, it replaces it, and the procedure is repeated. Otherwise the incumbent is "locally optimal" with respect to the neighborhood, and the heuristic terminates. Again we present two examples.

Example 2.6 Uncapacitated Facility Location. Consider an instance with $m = 6$ clients and $n = 4$ depots, and costs as shown below:

$$(c_{ij}) = \begin{matrix} & 1 & 2 & 3 & 4 \\ 1 & 6 & 2 & 3 & 4 \\ 2 & 1 & 9 & 4 & 11 \\ 3 & 15 & 2 & 6 & 3 \\ 4 & 9 & 11 & 4 & 8 \\ 5 & 7 & 23 & 2 & 9 \\ 6 & 4 & 3 & 1 & 5 \end{matrix} \quad \text{and } f_j = (21, 16, 11, 24).$$

← depots

↑ clients

↑ fixed cost of using facility j

Note that if $N = \{1,2,3,4\}$ denotes the set of depots, and $S \subseteq N$ the set of open depots, the associated cost is

(margin note: Since this is the uncapacitated problem, we service client i from the min cost facility j in our set S)

$$c(S) = \sum_{i=1}^{6} \min_{j \in S} c_{ij} + \sum_{j \in S} f_j.$$

Thus if $S^0 = \{1,2\}$ is the initial incumbent, $c(S^0) = (2+1+2+9+7+3) + 21 + 16 = 61$.

Now we need to define a neighborhood of S. One possibility is to consider as neighbors all sets obtained from S by the addition or removal of a single element:

(margin note: add an element or remove an element)

$$Q(S) = \{T \subseteq N : T = S \cup \{j\} \text{ for } j \notin S \text{ or } T = S \setminus \{i\} \text{ for } i \in S\}.$$

Thus $Q(S^0) = \{\{1\}, \{2\}, \{1,2,3\}, \{1,2,4\}\}$ with costs $c(1) = 63, c(2) = 66$, $c(123) = 60, c(124) = 84$.

The new incumbent is $S^1 = \{123\}$ with $c(S^1) = 60$, and $Q(S^1) = \{\{1,2\}, \{1,3\}, \{2,3\}, \{1,2,3,4\}\}$.

The new incumbent is $S^2 = \{23\}$ with $c(S^2) = 42$, and $Q(S^2) = \{\{2\}, \{3\}, \{1,2,3\}, \{2,3,4\}\}$.

The new incumbent is $S^3 = \{3\}$ with $c(S^3) = 31$, and $Q(S^3) = \{\{1,3\}, \{2,3\}, \{3,4\}, \phi\}$.

There is no improvement, so $S^3 = \{3\}$ is a locally optimal solution. ∎

Example 2.7 The Graph Equipartition Problem. Given a graph $G = (V,E)$ and $n = |V|$, the problem is to find a subset of nodes $S \subset V$ with $|S| = \lfloor \frac{n}{2} \rfloor$, for which the number of edges in the cutset $\delta(S, V \setminus S)$ is minimized, where $\delta(S, V \setminus S) = \{(i,j) \in E : i \in S, j \in V \setminus S\}$.

Again we need to define a neighborhood. As the feasible sets S are all of the same size, one possibility is to consider as neighbors all sets obtained by replacing one element in S by one element not in S:

$$Q(S) = \{T \subset V : |T \setminus S| = |S \setminus T| = 1\}.$$

We consider an instance on 6 nodes with edges $\{(1,4),(1,6),(2,3),(2,5), (2,6),(3,4),(3,5),(4,6)\}$.

Starting with $S^0 = \{1,2,3\}$, $c(S^0) = |(\delta(S^0, V \setminus S^0)| = 6$ as $\delta(S^0, V \setminus S^0) = \{(14),(16),(25),(26),(34),(35)\}$.

Here $Q(S^0) = \{(124),(125),(126),(134),(135),(136),(234),(235),(236)\}$ with $c(T) = 6,5,4,4,5,6,5,2,5$ respectively.

The new incumbent is $S^1 = \{235\}$ with $c(S^1) = 2$.

$Q(S^1)$ does not contain a better solution, and so S^1 is locally optimal. ∎

Other ways to generate primal feasible solutions are investigated in Chapter 12.

2.7 NOTES

2.1 The concept of a relaxation, and the complementary idea of a restriction are formalized in [GeoMar72].

2.2 Linear programming relaxations have been present ever since combinatorial problems were first formulated as linear integer programs [DanFulJoh54], [Dan57].

2.3 The assignment relaxation for the traveling salesman problem was already used in [Litetal63], and the 1-tree relaxation was introduced in [HelKar70].

2.4 Chapter 10 is on Lagrangian relaxation. Other relaxations studied include group or modular relaxations [Gom65] and Ch. II.3 in [NemWol88], and surrogate relaxations [Glo68].

2.5 A strong duality for the general matching problem appears in the classic paper [Edm65b], which has had a major influence on the development of combinatorial optimization. Several of the most beautiful results in this field are strong duality theorems; see [CooCunetal97]. For integer programs a general superadditive duality theory was developed in the seventies based on the work of [Gom69] and [GomJoh72]. See Ch. II.1 in [NemWol88].

2.6 The greedy and local exchange heuristics are formalized in Chapter 12, and other heuristic approaches are presented.

2.8 EXERCISES

1. Find a maximum cardinality matching in the graph of Figure 2.4 by inspection. Give a proof that the solution found is optimal.

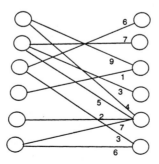

Fig. 2.4 Matching instance

2. A *stable set* is a set of nodes $U \subseteq V$ such that there are no edges between any two nodes of U. A *clique* is a set of nodes $U \subseteq V$ such that there is an edge between every pair of nodes in U. Show that the problem of finding a maximum cardinality stable set is dual to the problem of finding a minimum cover of the edges by cliques. Use this observation to find bounds on the maximum size of a stable set in the graph shown in Figure 2.5.

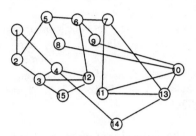

Fig. 2.5 Stable set instance

3. Find primal and dual bounds for the integer knapsack problem:

$$z = \max 42x_1 + 26x_2 + 35x_3 + 71x_4 + 53x_5$$
$$14x_1 + 10x_2 + 12x_3 + 25x_4 + 20x_5 \leq 69$$
$$x \in Z_+^5.$$

4. Consider the 0–1 integer program:

$$(P_1) \qquad \max\{cx : \sum_{j=1}^{n} a_{ij}x_j = b_i \text{ for } i = 1, \ldots, m, x \in B^n\},$$

and the 0–1 equality knapsack problem

$$(P_2) \qquad \max\{cx : \sum_{j=1}^{n}(\sum_{i=1}^{m} u_i a_{ij})x_j = \sum_{i=1}^{m} u_i b_i, x \in B^n\},$$

where $u \in R^m$. Show that P_2 is a relaxation of P_1.

5. Consider the equality integer knapsack problem:

$$(P_1). \qquad \min\{\sum_{j=1}^{5} c_j x_j : \frac{7}{4}x_1 - \frac{2}{3}x_2 + \frac{5}{2}x_3 - \frac{5}{12}x_4 + \frac{19}{6}x_5 = \frac{8}{3}, x \in Z_+^5\}$$

(i) Show that the problem

$$(P_2) \quad \min\{\sum_{j=1}^{5} c_j x_j : \frac{3}{4}x_1 + \frac{1}{3}x_2 + \frac{1}{2}x_3 + \frac{7}{12}x_4 + \frac{1}{6}x_5 = \frac{2}{3} + w, x \in Z_+^5, w \in Z_+^1\}$$

is a relaxation of P_1.

(ii) Show that the problem

$$(P_3) \quad \min\{\sum_{j=1}^{5} c_j x_j : \frac{3}{4}x_1 + \frac{1}{3}x_2 + \frac{1}{2}x_3 + \frac{7}{12}x_4 + \frac{1}{6}x_5 \geq \frac{2}{3}, x \in R_+^5\}$$

is a relaxation of P_2.

6. Consider a directed graph $D = (V, A)$ with arc lengths $c_e \geq 0$ for $e \in A$. Taking two distinct nodes $s, t \in V$, consider the problem of finding a shortest path from s to t. Show that

$$\max\{\pi_t : \pi_j - \pi_i \leq c_{ij} \text{ for } e = (i,j) \in A, \pi \in R_+^{|V|}, \pi_s = 0\}$$

is a strong dual problem.

7. Apply a greedy heuristic to the instance of the uncapacitated facility location problem in Section 2.6.

8. Define greedy and local search heuristics for the maximum cardinality stable set problem discussed in Exercise 2.2.

9. Formulate the maximum cardinality matching problem of Figure 2.4 as an integer program and solve its linear programming relaxation. Find a maximum weight matching with the weights shown. Check that the linear programming relaxation is again integral.

3
Well-Solved Problems

3.1 PROPERTIES OF EASY PROBLEMS

Here we plan to study some integer and combinatorial optimization problems that are "well-solved" in the sense that an "efficient" algorithm is known for solving all instances of the problem. Clearly an instance with 1000 variables or data values ranging up to 10^{20} can be expected to take longer than an instance with 10 variables and integer data never exceeding 100. So we need to define what we mean by efficient.

For the moment we will be very imprecise and say that an algorithm on a graph $G = (V, E)$ with n nodes and m edges is *efficient* if, in the worst case, the algorithm requires $0(m^p)$ elementary calculations (such as additions, divisions, comparisons, etc) for some integer p, where we assume that $m \geq n$.

In considering the COP $\max\{cx : x \in X \subseteq R^n\}$, it is not just of interest to find a dual problem, but also to consider a related problem, called the separation problem.

Definition 3.1 The *Separation Problem* associated with COP is the problem: Given $x^* \in R^n$, is $x^* \in conv(X)$? If not, find an inequality $\pi x \leq \pi_0$ satisfied by all points in X, but violated by the point x^*.

Now, in examining a problem to see if it has an efficient algorithm, we will see that the following four properties often go together:

(i) *Efficient Optimization Property:* For a given class of optimization problems (P) $\max\{cx : x \in X \subseteq R^n\}$, there exists an efficient (polynomial) algorithm.

(ii) *Strong Dual Property:* For the given problem class, there exists a strong dual problem (D) $\min\{\omega(u) : u \in U\}$ allowing us to obtain optimality conditions that can be quickly verified:

$x^* \in X$ is optimal in P if and only if there exists $u^* \in U$ with $cx^* = \omega(u^*)$.

(iii) *Efficient Separation Property:* There exists an efficient algorithm for the separation problem associated with the problem class.

(iv) *Explicit Convex Hull Property:* A compact description of the convex hull $conv(X)$ is known, which in principle allows us to replace every instance by the linear program: $\max\{cx : x \in conv(X)\}$.

Note that if a problem has the Explicit Convex Hull Property, then the dual of the linear program $\max\{cx : x \in conv(X)\}$ suggests that the Strong Dual Property should hold, and also using the description of $conv(X)$, there is some likelihood that the Efficient Separation Property holds. So some ties between the four properties are not surprising. The precise relationship will be discussed later. In the next sections we examine several classes of problems for which we will see that typically all four properties hold.

3.2 IPS WITH TOTALLY UNIMODULAR MATRICES

A natural starting point in solving integer programs :

$$(IP) \qquad \max\{cx : Ax \leq b, x \in Z_+^n\}$$

with integral data (A, b) is to ask when one will be lucky, and the linear programming relaxation (LP) $\max\{cx : Ax \leq b, x \in R_+^n\}$ will have an optimal solution that is integral.

From linear programming theory, we know that basic feasible solutions take the form: $x = (x_B, x_N) = (B^{-1}b, 0)$ where B is an $m \times m$ nonsingular submatrix of (A, I) and I is an $m \times m$ identity matrix.

Observation 3.1 (Sufficient Condition) If the optimal basis B has $det(B) = \pm 1$, then the linear programming relaxation solves IP.

Proof. From Cramer's rule, $B^{-1} = B^*/det(B)$ where B^* is the adjoint matrix. The entries of B^* are all products of terms of B. Thus B^* is an integral matrix, and as $det(B) = \pm 1$, B^{-1} is also integral. Thus $B^{-1}b$ is integral for all integral b. ∎

The next step is to ask when one will always be lucky. When do all bases or all optimal bases satisfy $det(B) = \pm 1$?

Definition 3.2 A matrix A is *totally unimodular (TU)* if every square submatrix of A has determinant $+1, -1$ or 0.

$$\begin{pmatrix} 1 & -1 \\ 1 & 1 \end{pmatrix} \qquad \begin{pmatrix} 1 & 1 & 0 \\ 0 & 1 & 1 \\ 1 & 0 & 1 \end{pmatrix}$$

Table 3.1   Matrices that are not TU

$$\begin{pmatrix} 1 & -1 & -1 & 0 \\ -1 & 0 & 0 & 1 \\ 0 & 1 & 0 & -1 \\ 0 & 0 & 1 & 0 \end{pmatrix} \qquad \begin{pmatrix} 0 & 1 & 0 & 0 & 0 \\ 0 & 1 & 1 & 1 & 1 \\ 1 & 0 & 1 & 1 & 1 \\ 1 & 0 & 0 & 1 & 0 \\ 1 & 0 & 0 & 0 & 0 \end{pmatrix}$$

Table 3.2   Matrices that are TU

First we consider whether such matrices exist and how we can recognize them. Some simple observations follow directly from the definition.

Observation 3.2 If A is TU, $a_{ij} \in \{+1, -1, 0\}$ for all i, j.

Observation 3.3 The matrices in Table 3.1 are not TU. The matrices in Table 3.2 are TU.

Proposition 3.1 *A matrix A is TU if and only if*
(i) *the transpose matrix A^T is TU if and only if*
(ii) *the matrix (A, I) is TU.*

There is a simple and important sufficient condition for total unimodularity, that can be used to show that the first matrix in Table 3.2 is TU.

Proposition 3.2 *(Sufficient Condition). A matrix A is TU if*
(i) $a_{ij} \in \{+1, -1, 0\}$ *for all i, j.*
(ii) *Each column contains at most two nonzero coefficients ($\sum_{i=1}^m |a_{ij}| \leq 2$).*
(iii) *There exists a partition (M_1, M_2) of the set M of rows such that each column j containing two nonzero coefficients satisfies $\sum_{i \in M_1} a_{ij} - \sum_{i \in M_2} a_{ij} = 0$.*

Proof. Assume that A is not TU, and let B be the smallest square submatrix of A for which $\det(A) \notin \{0, 1, -1\}$. B cannot contain a column with a single nonzero entry, as otherwise B would not be minimal. So B contains two nonzero entries in each column. Now by condition (iii), adding the rows in M_1 and subtracting the rows in M_2 gives the zero vector, and so $\det(B) = 0$, and we have a contradiction. ∎

Note that condition (iii) means that if the nonzeros are in rows i and k, and if $a_{ij} = -a_{kj}$, then $\{i, k\} \in M_1$ or $\{i, k\} \in M_2$, whereas if $a_{ij} = a_{kj}$, $i \in M_1$

and $k \in M_2$, or vice versa. This leads to a simple algorithm to test whether the conditions of Proposition 3.2 hold. In the next section we will see an important class of matrices arising from network flow problems that satisfy this sufficient condition.

Now returning to IP, it is clear that when A is TU, the linear programming relaxation solves IP. In some sense the converse holds.

Proposition 3.3 *The linear program* $\max\{cx : Ax \leq b, x \in R^n_+\}$ *has an integral optimal solution for all integer vectors b for which it has a finite optimal value if and only if A is totally unimodular.*

On the question of efficient algorithms, we have essentially proved that for the IP: $\max\{cx : Ax \leq b, x \in Z^n_+\}$ with A totally unimodular:

(a) The Strong Dual Property holds: the linear program $(D) : \min\{ub : uA \geq c, u \geq 0\}$ is a strong dual.
(b) The Explicit Convex Hull Property holds: the convex hull of the set of feasible solutions $conv(X) = \{Ax \leq b, x \geq 0\}$ is known.
(c) The Efficient Separation Property holds: the separation problem is easy as it suffices to check if $Ax^* \leq b$ and $x^* \geq 0$. (since we already have the convex hull, all we really need to check is feasibility of the LP)

Given that these three properties hold, we have suggested that the Efficient Optimization Property should also hold, so there should be an efficient algorithm for IP. This turns out to be true, but it is a nontrivial result beyond the scope of this text. This is in turn related to the fact that efficient algorithms to recognize whether a matrix A is TU are also nontrivial.

3.3 MINIMUM COST NETWORK FLOWS

Here we consider an important class of problems with many applications lying at the frontier between linear and integer programming.

Given a digraph $D = (V, A)$ with arc capacities h_{ij} for all $(i, j) \in A$, demands b_i (positive inflows or negative outflows) at each node $i \in V$, and unit flow costs c_{ij} for all $(i, j) \in A$, the minimum cost network flow problem is to find a feasible flow that satisfies all the demands at minimum cost. This has the formulation:

$$\min \sum_{(i,j) \in A} c_{ij} x_{ij} \tag{3.1}$$

$$\sum_{k \in V^+(i)} x_{ik} - \sum_{k \in V^-(i)} x_{ki} = b_i \text{ for } i \in V \tag{3.2}$$

$$0 \leq x_{ij} \leq h_{ij} \quad \text{for } (i, j) \in A \tag{3.3}$$

where x_{ij} denotes the flow in arc (i, j), $V^+(i) = \{k : (i, k) \in A\}$ and $V^-(i) = \{k : (k, i) \in A\}$.

It is evident that for the problem to be feasible the total sum of all the demands must be zero (i.e., $\sum_{i \in V} b_i = 0$).

Example 3.1 The digraph in Figure 3.1 leads to the following set of balance

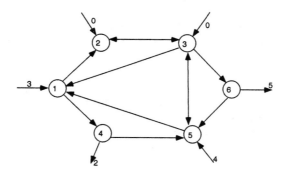

Fig. 3.1 Digraph for minimum cost network flow

equations:

x_{12}	x_{14}	x_{23}	x_{31}	x_{32}	x_{35}	x_{36}	x_{45}	x_{51}	x_{53}	x_{65}		
1	1	0	−1	0	0	0	0	−1	0	0	=	3
−1	0	1	0	−1	0	0	0	0	0	0	=	0
0	0	−1	1	1	1	1	0	0	−1	0	=	0
0	−1	0	0	0	0	0	1	0	0	0	=	−2
0	0	0	0	0	−1	0	−1	1	1	−1	=	4
0	0	0	0	0	0	−1	0	0	0	1	=	−5

The additional constraints are the bound constraints: $0 \leq x_{ij} \leq h_{ij}$. ∎

Proposition 3.4 *The constraint matrix A arising in a minimum cost network flow problem is totally unimodular.*

Proof. The matrix A is of the form $\begin{pmatrix} C \\ I \end{pmatrix}$ where C comes from the flow conservation constraints, and I from the upper bound constraints. Therefore it suffices to show that C is TU. The sufficient conditions of Proposition 3.2 are satisfied with $M_1 = M$ and $M_2 = \phi$. ∎

Corollary In a minimum cost network flow problem, if the demands $\{b_i\}$ and the capacities $\{h_{ij}\}$ are integral,
(i) Each extreme point is integral.
(ii) The constraints (3.2)-(3.3) describe the convex hull of the integral feasible flows.

3.4 SPECIAL MINIMUM COST FLOWS

The Shortest Path Problem. Given a digraph $D = (V, A)$, two distinguished nodes $s, t \in V$, and nonnegative arc costs c_{ij} for $(i, j) \in A$, find a minimum cost $s - t$ path.

The Max Flow Problem. Given a digraph $D = (V, A)$, two distinguished nodes $s, t \in V$, and nonnegative capacities h_{ij} for $(i, j) \in A$, find a maximum flow from s to t.

Both these problems are special cases of the minimum cost network flow problem so we can use total unimodularity to analyze them. However, the reader has probably already seen combinatorial polynomial algorithms for these two problems in a course on network flows.

What are the associated dual problems?

3.4.1 Shortest Path

Observe first that the shortest path problem can be formulated as:

$$z = \min \sum_{(i,j) \in A} c_{ij} x_{ij} \tag{3.4}$$

$$\sum_{k \in V^+(i)} x_{ik} - \sum_{k \in V^-(i)} x_{ki} = 1 \text{ for } i = s \tag{3.5}$$

$$\sum_{k \in V^+(i)} x_{ik} - \sum_{k \in V^-(i)} x_{ki} = 0 \text{ for } i \in V \setminus \{s, t\} \tag{3.6}$$

$$\sum_{k \in V^+(i)} x_{ik} - \sum_{k \in V^-(i)} x_{ki} = -1 \text{ for } i = t \tag{3.7}$$

$$x_{ij} \geq 0 \text{ for } (i, j) \in A \tag{3.8}$$

$$x \in Z^{|A|} \tag{3.9}$$

where $x_{ij} = 1$ if arc (i, j) is in the minimum cost (shortest) $s - t$ path.

Theorem 3.5 z is the length of a shortest $s - t$ path if and only if there exist values π_i for $i \in V$ such that $\pi_s = 0, \pi_t = z$, and $\pi_j - \pi_i \leq c_{ij}$ for $(i, j) \in A$.

Proof. The linear programming dual of (3.4)–(3.8) is precisely

$$w^{LP} = \max \pi_t - \pi_s$$
$$\pi_j - \pi_i \leq c_{ij} \text{ for } (i, j) \in A.$$

Replacing π_j by $\pi_j + \alpha$ for all $j \in V$ does not change the dual, so we can fix $\pi_s = 0$ without loss of generality. As the primal matrix is totally unimodular, strong duality holds and the claim follows. ∎

We note that one particular dual solution is obtained by taking π_i to be the cost of a shortest path from s to i.

3.4.2 Maximum $s - t$ Flow

Adding a backward arc from t to s, the maximum $s - t$ flow problem can be formulated as:

$$\max x_{ts}$$
$$\sum_{k \in V^+(i)} x_{ik} - \sum_{k \in V^-(i)} x_{ki} = 0 \text{ for } i \in V$$
$$0 \le x_{ij} \le h_{ij} \text{ for } (i,j) \in A.$$

The dual is:

$$\min \sum_{(i,j) \in A} h_{ij} w_{ij}$$
$$u_i - u_j + w_{ij} \ge 0 \text{ for } (i,j) \in A$$
$$u_t - u_s \ge 1.$$

From total unimodularity, an optimal solution is integer. Also as the dual is unchanged if we replace u_j by $u_j + \alpha$ for all $j \in V$, we can set $u_s = 0$. Given such a solution, let $X = \{j \in V : u_j \le 0\}$ and $\bar{X} = V \setminus X = \{j \in V : u_j \ge 1\}$. Now

$$\sum_{(i,j) \in A} h_{ij} w_{ij} \ge \sum_{(i,j) \in A, i \in X, j \in \bar{X}} h_{ij} w_{ij} \ge \sum_{(i,j) \in A, i \in X, j \in \bar{X}} h_{ij}$$

as $w_{ij} \ge u_j - u_i \ge 1$ for $(i,j) \in A$ with $i \in X$ and $j \in \bar{X}$.

However, this lower bound $\sum_{(i,j) \in A, i \in X, j \in \bar{X}} h_{ij}$ is attained by the solution $u_j = 0$ for $j \in X$, $u_j = 1$ for $j \in \bar{X}$, $w_{ij} = 1$ for $(i,j) \in A$ with $i \in X$ and $j \in \bar{X}$, and $w_{ij} = 0$ otherwise. So there is an optimal 0-1 solution.

We see that $s \in X, t \in \bar{X}$, $\{(i,j) : w_{ij} = 1\}$ is the set of arcs of the $s - t$ cut $(X, V \setminus X)$, and we obtain the standard result that the maximum value of an $s - t$ flow equals the minimum capacity of an $s - t$ cut.

Theorem 3.6 *A strong dual to the max $s - t$ flow problem is the minimum $s - t$ cut problem:*

$$\min_{X} \{ \sum_{(i,j) \in A : i \in X, j \notin X} h_{ij} : s \in X \subset V \setminus \{t\} \}.$$

3.5 OPTIMAL TREES

Definition 3.3 *Given a graph $G = (V, E)$, a* forest *is a subgraph $G' = (V, E')$ containing no cycles.*

Definition 3.4 *Given a graph $G = (V, E)$, a* tree *is a subgraph $G' = (V, E')$ that is a forest and is* connected *(contains a path between every pair of nodes of V).*

44 WELL-SOLVED PROBLEMS

Some well-known consequences of these definitions are listed below:

Proposition 3.7 *A graph $G = (V, E)$ is a tree if and only if*
(i) it is a forest containing exactly $n - 1$ edges, if and only if
(ii) it is an edge-minimal connected graph spanning V, if and only if
(iii) it contains a unique path between every pair of nodes of V, if and only if
(iv) the addition of an edge not in E creates a unique cycle.

The Maximum Weight Forest (Tree) Problem. Given a graph $G = (V, E)$ and edge weights c_e for $e \in E$, find a maximum weight subgraph that is a forest (tree).

This problem arises naturally in many telecommunications and computer network applications where it is necessary that there is at least one path between each pair of nodes. When one wishes to minimize installation costs, the optimal solution is clearly a minimum weight tree.

Remember from the previous chapter that the idea of a *greedy algorithm* is to take the best element and run. It is very shortsighted. It just chooses one after the other whichever element gives the maximum profit and still gives a feasible solution. For a graph $G = (V, E)$, we let $n = |V|$ and $m = |E|$.

Greedy Algorithm for a Maximum Weight Tree

Initialization. Set $E^0 = E, T^0 = \phi$. Order the edges by nonincreasing weight $c_1 \geq c_2 \geq \ldots \geq c_m$, where c_t is the cost of edge e_t.
Iteration t. If $T^{t-1} \cup \{e_t\}$ contains no cycle, set $T^t = T^{t-1} \cup \{e_t\}$. Otherwise $T^t = T^{t-1}$.

Set $E^t = E^{t-1} \setminus \{e_t\}$. If $|T^t| = n - 1$, stop with T^t is optimal. If $t = m$, stop with no feasible solution.

To obtain a maximum weight **forest**, it suffices to modify the greedy algorithm to stop as soon as $c_{t+1} \leq 0$.

Example 3.2 Consider the graph shown in Figure 3.2 with the weights on the edges as shown.

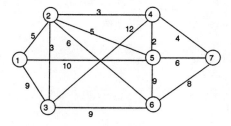

Fig. 3.2 Graph for optimal weight tree

We consider the edges in the order:

	e_1	e_2	e_3	e_4	e_5	e_6	e_7
(i,j)	$(3,4)$	$(1,5)$	$(1,3)$	$(3,6)$	$(5,6)$	$(6,7)$	$(5,7)$
c_e	12	10	9	9	9	8	6
	+	+	+	+		+	

	e_8	e_9	e_{10}	e_{11}	e_{12}	e_{13}	e_{14}
(i,j)	$(2,6)$	$(1,2)$	$(2,5)$	$(4,7)$	$(2,3)$	$(2,4)$	$(4,5)$
c_e	6	5	5	4	3	3	2
		+					

The algorithm chooses the edges marked with a + and rejects the others. For example, the first edge rejected is $e_5 = (5,6)$ because it forms a cycle with e_2, e_3, and e_4, which have already been selected. ∎

Theorem 3.8 *The greedy algorithm terminates with an optimal weight tree.*

Proof. Suppose for simplicity that all the edge weights are different. Let $T = \{g_1, \ldots, g_{n-1}\}$ be the edges chosen by the greedy algorithm with $c_{g_1} > \cdots > c_{g_{n-1}}$. Let $F = \{f_1, \ldots, f_{n-1}\}$ be the edges of an optimal solution with $c_{f_1} > \cdots > c_{f_{n-1}}$.

If the solutions are the same, the result is proved. So suppose the two solutions differ with $g_1 = f_1, \ldots, g_{k-1} = f_{k-1}$ but $g_k \neq f_k$.

(i) We observe that $c_{g_k} > c_{f_k}$ because the greedy algorithm chooses g_k and not f_k and neither edge creates a cycle with $\{g_1, \ldots, g_{k-1}\}$. Also by construction $c_{f_1} > \cdots > c_{f_{n-1}}$, and so $g_k \notin F$.

(ii) Now consider the edge set $F \cup \{g_k\}$. As F is a tree, it follows from Proposition 3.7 that $F \cup \{g_k\}$ contains exactly one cycle C. Note however that the set of edges $\{f_1, \ldots, f_{k-1}, g_k\} = \{g_1, \ldots, g_{k-1}, g_k\}$ forms part of the tree T and thus does not contain a cycle. Therefore one of the other edges $f_k, \ldots, f_{n-1}$ must be in the cycle C. Suppose f^* is one such edge.

(iii) As C contains a unique cycle and f^* is in this cycle, $T' = F \cup \{g_k\} \setminus \{f^*\}$ is cycle free. As it has $n-1$ edges, it is a tree.

(iv) Finally, as $c_{g_k} > c_{f_k}$ and $c_{f_k} \geq c_{f^*}$, the weight of T' exceeds that of F.

(v) As F is an optimal tree, we have arrived at a contradiction, and so T and F cannot differ. ∎

As the greedy algorithm is easily seen to be polynomial, the Efficient Optimization Property holds for the maximum weight tree problem. So we can again consider the other three properties. To do this, we need a formulation of the problem as an integer program. In modeling the traveling salesman problem in Section 1.2 we saw how to avoid cycles. Thus the maximum weight forest problem can be formulated as:

$$\max \sum_{e \in E} c_e x_e \tag{3.10}$$
$$\sum_{e \in E(S)} x_e \leq |S| - 1 \text{ for } 2 \leq |S| \leq n \tag{3.11}$$

$$x_e \geq 0 \text{ for } e \in E \qquad (3.12)$$
$$x \in Z^{|E|}. \qquad (3.13)$$

Theorem 3.9 *The convex hull of the incidence vectors of the forests in a graph is given by the constraints (3.11)–(3.12).*

This result says that the Explicit Convex Hull Property holds for the Maximum Weight Forest Problem. We generalize and prove this result in the next section, and in Chapter 9 we show that the Efficient Separation Property holds for this problem.

To terminate this section we introduce an important and more difficult generalization of the optimal tree problem. Given a graph $G = (V, E)$ and a set of terminals $T \subseteq V$, a *Steiner tree on* T is an edge-minimal acyclic subgraph of G containing a path joining every pair of nodes in T. Such a subgraph may or may not have edges incident to the nodes in $V \setminus T$. Given weights c_e for $e \in E$, the *Optimal Steiner Tree Problem* is to find a minimum weight Steiner tree. Observe that when $T = V$, this is the optimal tree problem, and when $|T| = 2$, it is the shortest path problem.

3.6 SUBMODULARITY AND MATROIDS*

Here we examine a larger class of problems for which a greedy algorithm provides an optimal solution. This generalizes the maximum weight forest problem examined in the last section. $\mathcal{P}(N)$ denotes the set of subsets of N.

Definition 3.5 (i) A set function $f : \mathcal{P}(N) \to R^1$ is *submodular* if
$$f(A) + f(B) \geq f(A \cap B) + f(A \cup B) \text{ for all } A, B \subseteq N.$$

(ii) A set function f is *nondecreasing* if
$$f(A) \leq f(B) \text{ for all } A, B \text{ with } A \subset B \subseteq N.$$

An alternative representation of such functions is useful.

Proposition 3.10 *A set function f is non-decreasing and submodular if and only if*
$$f(A) \leq f(B) + \sum_{j \in A \setminus B} [f(B \cup \{j\}) - f(B)] \text{ for all } A, B \subseteq N.$$

Proof. Suppose f is non-decreasing and submodular. Let $A \setminus B = \{j_1, \ldots, j_r\}$. Then $f(A) \leq f(A \cup B) = \sum_{i=1}^{r}[f(B \cup \{j_1, \ldots, j_i\}) - f(B \cup \{j_1, \ldots, j_{i-1}\})] \leq$

$\sum_{i=1}^{r}[f(B \cup \{j_i\}) - f(B)] = \sum_{j \in A \setminus B}[f(B \cup \{j\}) - f(B)]$, where the first inequality follows from f nondecreasing and the second from submodularity. The other direction is immediate. ∎

Now given a nondecreasing submodular function f on N with $f(\emptyset) = 0$, we consider the *submodular polyhedron*:

$$P(f) = \{x \in R_+^n : \sum_{j \in S} x_j \leq f(S) \text{ for } S \subseteq N\},$$

and the associated *submodular optimization problem*:

$$\max\{cx : x \in P(f)\}.$$

The Greedy Algorithm for the Submodular Optimization Problem

(i) Order the variables so that $c_1 \geq c_2 \geq \ldots \geq c_r > 0 \geq c_{r+1} \geq \ldots \geq c_n$.
(ii) Set $x_i = f(S^i) - f(S^{i-1})$ for $i = 1, \ldots, r$ and $x_j = 0$ for $j > r$, where $S^i = \{1, \ldots, i\}$ for $i = 1, \ldots, r$ and $S^0 = \emptyset$.

Theorem 3.11 *The greedy algorithm solves the submodular optimization problem.*

Proof. As f is nondecreasing, $x_i = f(S^i) - f(S^{i-1}) \geq 0$ for $i = 1, \ldots, r$. Also for each $T \subseteq N$,

$$\begin{aligned}
\sum_{j \in T} x_j &= \sum_{j \in T \cap S^r} [f(S^j) - f(S^{j-1})] \\
&\leq \sum_{j \in T \cap S^r} [f(S^j \cap T) - f(S^{j-1} \cap T)] \\
&\leq \sum_{j \in S^r} [f(S^j \cap T) - f(S^{j-1} \cap T)] \\
&= f(S^r \cap T) - f(\emptyset) \leq f(T),
\end{aligned}$$

where the first inequality follows from the submodularity of f, and the others as f is nondecreasing. So the greedy solution is feasible with value $\sum_{i=1}^{r} c_i[f(S^i) - f(S^{i-1})]$.

Now consider the linear programming dual:

$$\min \sum_{S \subseteq N} f(S) y_S$$
$$\sum_{S : j \in S} y_S \geq c_j \text{ for } j \in N$$
$$y_S \geq 0 \text{ for } S \subseteq N.$$

Let $y_{S^i} = c_i - c_{i+1}$ for $i = 1, \ldots, r-1$, $y_{S^r} = c_r$, and $y_S = 0$ otherwise. Clearly $y_S \geq 0$ for all $S \subseteq N$. Also for $j \leq r$, $\sum_{S : j \in S} y_S \geq \sum_{i=j}^{r} y_{S^i} =$

$\sum_{i=j}^{r-1}(c_i - c_{i+1}) + c_r = c_j$, and for $j > r$, $\sum_{S:j \in S} y_S \geq 0 \geq c_j$. Thus the solution y is dual feasible. Finally the dual objective value is

$$\sum_{i=1}^{r} f(S^i) y_{S^i} = \sum_{i=1}^{r-1} f(S^i)(c_i - c_{i+1}) + f(S^r) c_r = \sum_{i=1}^{r} c_i [f(S^i) - f(S^{i-1})].$$

So the value of the dual feasible solution has the same value as the greedy solution, and so from linear programming duality, the greedy solution is optimal. ∎

Note that when f is integer-valued, the greedy algorithm provides an integral solution. In the special case when $f(S \cup \{j\}) - f(S) \in \{0, 1\}$ for all $S \subset N$ and $j \in N \setminus S$, we call f a *submodular rank function*, and the greedy solution is a $0 - -1$ vector. What is more, we now show that the feasible 0–1 points in the submodular rank polyhedron generate an interesting combinatorial structure, called a *matroid*.

Proposition 3.12 *Suppose that r is a submodular rank function on a set N with $r(\emptyset) = 0$.*
(i) $r(A) \leq |A|$ for all $A \subseteq N$.
(ii) If $r(A) = |A|$, then $r(B) = |B|$ for all $B \subset A \subseteq N$.
(iii) If x^A is the incidence vector of $A \subseteq N$, $x^A \in P(r)$ if and only if $r(A) = |A|$.

Proof. (i) Using Proposition 3.10 and the property of a submodular rank function, $r(A) \leq r(\emptyset) + \sum_{j \in A}[r(\{j\}) - r(\emptyset)] \leq |A|$.
(ii) Again using the same properties, $|A| = r(A) \leq r(B) + \sum_{j \in A \setminus B}[r(B \cup \{j\}) - r(B)] \leq |B| + |A \setminus B| = |A|$. Equality must hold throughout, and thus $r(B) = |B|$.
(iii) If $r(A) < |A|$, $\sum_{j \in A} x_j^A = |A| > r(A)$ and $x^A \notin P(r)$. If $r(A) = |A|$, $\sum_{j \in S} x_j^A = |A \cap S| = r(A \cap S) \leq r(S)$ where the second equality uses (ii). This inequality holds for all $S \subseteq N$, and thus $x^A \in P(r)$. ∎

Definition 3.6 *Given a submodular rank function r, a set $A \subseteq N$ is independent if $r(A) = |A|$. The pair $(N, \mathcal{F})$, where $\mathcal{F}$ is the set of independent sets, is called a matroid.*

Based on Theorem 3.11, we know how to optimize on matroids.

Theorem 3.13 *The greedy algorithm solves the maximum weight independent set problem in a matroid.*

Given a connected graph $G = (V, E)$, it is not difficult to verify that the edge sets of forests form a matroid and that the function $r : \mathcal{P}(E) \to R^1$, where $r(E')$ is the size of the largest forest in (V, E'), is a submodular rank function. Specifically when $S \subseteq V$, $E' = E(S)$ and the subgraph $(S, E(S))$

is connected, we clearly have $r(E(S)) = |S| - 1$, so the forest polyhedron (3.11)-(3.12) is a special case of a submodular polyhedron.

The constraint set associated to a submodular polyhedron has another interesting property.

Definition 3.7 A set of linear inequalities $Ax \leq b$ is called *Totally Dual Integral (TDI)* if, for all $c \in Z^n$ for which the linear program $\max\{cx : Ax \leq b\}$ has a finite optimal value, the dual linear program

$$\min\{yb : yA = c, y \geq 0\}$$

has an optimal solution with y integral.

We have seen in the proof of Theorem 3.11 that the linear system $\{\sum_{j \in S} x_j \leq f(S)$ for $S \subseteq N, x_j \geq 0$ for $j \in N\}$ is TDI. Based on the following result, the TDI property provides another useful way of showing that certain linear programs always have integer solutions.

Theorem 3.14 If $Ax \leq b$ is TDI, b is an integer vector, and $P = \{x \in R^n : Ax \leq b\}$ has vertices, then all vertices of P are integral.

Note also that if A is a TU matrix, then $Ax \leq b$ is TDI.

3.7 NOTES

3.1 The theoretical importance of the separation problem is discussed at the end of Chapter 6. Its practical importance was brought out in the first computational studies using strong cutting planes; see Chapter 9.

3.2 Totally unimodular matrices have been studied since the fifties. The characterization of Proposition 3.1 is due to [HofKru56]. The interval or consecutive 1's property of Exercise 3.3 is due to [FulGro65], and the stronger necessary condition is from [Gho62]. A complete characterization of TU matrices is much more difficult; see [Sey80] or the presentation in [Sch86].

3.3 We again refer to [AhuMagOrl93] for network flows, as well as shortest path and max flow problems. This book also contains a large number of applications and a wealth of exercises. Note also the chapter of Ahuja on flows and paths in [DelAMafMar97], which also indicates where some of the latest software for network flow problems can be obtained.

3.4 The max flow min cut theorem was already part of the max flow algorithm of [ForFul56]. The min cut problem arises as a separation problem in solving TSP and other network design problems. The problem of finding all minimum cuts was answered in [GomHu61]. Recently new algorithms have appeared that find minimum cuts directly without using flows; see Ch. 3 in

[CooCunetal97].

3.5 The greedy algorithm for finding minimum weight trees is from [Kru56]. A faster classical algorithm is that of [Prim57]. Special algorithms based on Delaunay triangulations can be used for two-dimensional Euclidean problems, [PreSha85]. [Goe94] and [MagWol95] contain a discussion of many alternative formulations for tree and Steiner tree problems.

3.6 Submodular polyhedra and the greedy algorithm for matroids are studied in [Edm70] and [Edm71], see also [Law76]. [Wel76] is a book devoted to matroids. For total dual integrality, see [EdmGil77].

3.8 EXERCISES

1. Are the following matrices totally unimodular or not?

$$A_1 = \begin{pmatrix} 1 & 0 & 1 & 0 & 1 \\ 0 & 1 & 1 & 1 & 0 \\ 0 & 0 & 0 & 1 & 1 \\ 1 & 1 & 0 & 0 & 0 \end{pmatrix}, A_2 = \begin{pmatrix} -1 & & 1 & & -1 & & \\ & 1 & & 1 & 1 & & & 1 \\ -1 & 1 & & & & & & \\ & & 1 & & & & 1 & 1 \\ & & & 1 & & & & -1 \end{pmatrix}.$$

2. Prove that the polyhedron $P = \{(x_1, \ldots, x_m, y) \in R_+^{m+1} : y \leq 1, x_i \leq y$ for $i = 1, \ldots, m\}$ has integer vertices.

3. A 0–1 matrix B has the *consecutive 1's property* if for any column j, $b_{ij} = b_{i'j} = 1$ with $i < i'$ implies $b_{lj} = 1$ for $i < l < i'$.
   A more general sufficient condition for total unimodularity is: Matrix A is TU if

   (i) $a_{ij} \in \{+1, -1, 0\}$ for all i, j.
   (ii) For any subset M of the rows, there exists a partition (M_1, M_2) of M such that each column j satisfies

   $$|\sum_{i \in M_1} a_{ij} - \sum_{i \in M_2} a_{ij}| \leq 1.$$

   Use this to show that a matrix with the consecutive 1's property is TU.

4. Consider a scheduling model in which a machine can be switched on at most k times: $\sum_t z_t \leq k, z_t - y_t + y_{t-1} \geq 0, z_t \leq y_t, 0 \leq y_t, z_t \leq 1$ for all t, where $y_t = 1$ if the machine is on in period t, and $z_t = 1$ if it is switched on in period t. Show that the resulting matrix is TU.

5. Prove Proposition 3.3.

6. Use linear programming to find the length of a shortest path from node s to node t in the directed graph of Figure 3.3. Use an optimal dual solution to prove that your solution is optimal.

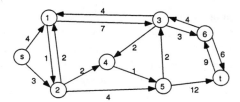

Fig. 3.3 Shortest path instance

7. Use linear programming to find a minimum $s - t$ cut in the capacitated network of Figure 3.4.

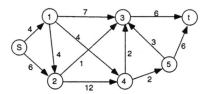

Fig. 3.4 Network instance

8. Find a minimum weight spanning tree in the graph shown in Figure 3.5.

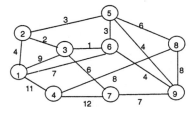

Fig. 3.5 Tree instance

9. Prove that the greedy algorithm produces an optimal weight tree when edge weights can be equal.

10. Formulate the optimal Steiner tree problem as an integer program.

11. (i) For their annual Christmas party the thirty members of staff of the thriving company Ipopt were invited/obliged to dine together and then spend the night in a fancy hotel. The boss's secretary had the unenviable task of allocating the staff two to a room. Knowing the likes and dislikes of everyone, she drew up a list of all the compatible pairs. How could you help her to fill all fifteen rooms?

(ii) Recently a summer camp was organized for an equal number of English and French children. After a few days, the children had to participate in an orienteering competition in pairs, each pair made up of one French and one English child. To allocate the pairs, each potential pair was asked to give a weight from 1 to 10 representing their willingness to form a pair. Formulate the problem of choosing the pairs so as to maximize the sum of the weights.

(iii) If you have a linear programming code available, can you help either the boss's secretary or the camp organizer or both?

12. Consider a real matrix C with n columns. Let $N = \{1, \ldots, n\}$ and $\mathcal{F} = \{S \subseteq N :$ the columns $\{c_j\}_{j \in S}$ are linearly independent$\}$. Show that $(N, \mathcal{F})$ is a matroid. What is the associated rank function r?

13. Given a matroid, show that
(i) if A and B are independent sets with $|A| > |B|$, then there exists $j \in A \setminus B$ such that $A \cup \{j\}$ is independent, and
(ii) for an arbitrary set $A \subseteq N$, every maximal independent set in A has the same cardinality.

4
Matchings and Assignments

4.1 AUGMENTING PATHS AND OPTIMALITY

Here we demonstrate two other important ideas used in certain combinatorial algorithms. One idea is that of a primal algorithm systematically moving from one feasible solution to a better one. The second is that of iterating between primal and dual problems using the LP complementarity conditions.

First a few reminders. We suppose that a graph $G = (V, E)$ is given.

Definition 4.1 A *matching* $M \subseteq E$ is a set of disjoint edges, that is, at most one edge of a matching is incident to any node $v \in V$.

Definition 4.2 A *covering by nodes* is a set of nodes $R \subseteq V$ such that every edge $e \in E$ is incident to at least one of the nodes of R.

We have shown in Section 2.5 that there is a weak duality between matchings and coverings by nodes, namely for every matching M and covering by nodes R, $|M| \leq |R|$. Here we consider the *Maximum Cardinality Matching Problem* $\max\{|M| : M \text{ is a matching}\}$, and to solve it we examine first how to construct matchings of larger and larger cardinality.

Definition 4.3 An *alternating path* with respect to a matching M is a path $P = v_0, e_1, v_1, e_2, \ldots, e_p, v_p$ such that
(i) $e_1, e_3, \ldots, e_{odd} \in E \setminus M$. $v_o - v_p$
(ii) $e_2, e_4, \ldots, e_{even} \in M$.

(iii) v_0 is not incident to the matching M (v_0 is an *exposed* node).

An *augmenting path* is an alternating path that in addition satisfies the condition:

(iv) The number of edges p is odd, and v_p is not incident to the matching M.

Augmenting paths are what we need (see Figure 4.1).

Proposition 4.1 *Given a matching M and an augmenting path P relative to M, the symmetric difference $M' = (M \cup P) \setminus (M \cap P)$ is a matching with $|M'| > |M|$.*

Proof. As v_0 and v_p do not touch M, M' is a matching. As p is odd, $|P \cap (E \setminus M)| = |P \cap M| + 1$. Thus $|M'| = |M| + 1$. ∎

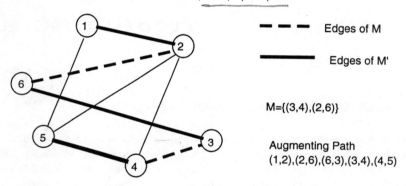

Fig. 4.1   An augmenting path

So the existence of an augmenting path implies that M is not optimal. Is the converse also true? If there is no augmenting path, can we conclude that M is optimal?

Proposition 4.2 *If there is no augmenting path relative to a matching M, then M is of maximum cardinality.*

Proof. We show the contrapositive. We suppose that M is not optimal. Thus there exists a matching M' with $|M'| > |M|$. Consider the graph whose edges are given by $(M \cup M') \setminus (M \cap M')$. The degree of each node in this graph is $0, 1$, or 2. Thus the connected components of the graph are paths and cycles. For a cycle C, the edges alternate between M and M', and so the cycles are of even length and contain the same number of edges from M and M'. The paths can contain either an even or an odd number of edges. As $|M'| > |M|$, one of the paths must contain more edges from M' than from M. This path is an augmenting path. ∎

Can we find whether there is an augmenting path or not in polynomial time? If there is no augmenting path, can we give a simple way to verify that the matching we have is maximum? In the next section we give a positive answer to both these questions when the graph is bipartite.

4.2 BIPARTITE MAXIMUM CARDINALITY MATCHING

Given a bipartite graph $G = (V_1, V_2, E)$, where $V = V_1 \cup V_2$ and every edge has one endpoint in V_1 and the other in V_2, we wish to find a maximum size matching in G. We suppose that a matching M (possibly empty) has been found, and we wish either to find an augmenting path, or to demonstrate that there is no such path and that the matching M is optimal.

We try to systematically examine all augmenting paths.

Observation 4.1 As augmenting paths P are of odd length and the graph is bipartite, one of the exposed nodes of P is in V_1 and the other in V_2. Thus it suffices to start enumerating from V_1.

Outline of the Algorithm. We start by labeling all the nodes of V_1 disjoint from M. These are candidates to be the first node of an alternating path.

The first (and subsequent odd) edges of an alternating path are in $E \setminus M$, and thus all such edges from the labeled nodes in V_1 are candidates. The endpoints of these edges in V_2 are then labeled.

The second (and subsequent even) edges of an alternating path are in M, and thus any edge in M touching a labeled node in V_2 is a candidate. The endpoints of these edges in V_1 are then labeled, and so on.

The labeling stops: either when a node of V_2 is labeled that is not incident to M, so an augmenting path has been found, or when no more edges can be labeled, and so none of the alternating paths can be extended further.

Algorithm for Bipartite Maximum Cardinality Matching

Step 0. $G = (V_1, V_2, E)$ is given. M is a matching. No nodes are labeled or scanned.

Step 1. (Labeling)

(1.0) Give the label $*$ to each exposed node in V_1.

(1.1) If there are no unscanned labels, go to Step 3. Choose a labeled unscanned node i. If $i \in V_1$, go to 1.2. If $i \in V_2$, go to 1.3.

(1.2) *Scan* the labeled node $i \in V_1$. For all $(i,j) \in E \setminus M$, give j the label i if j is unlabeled. Return to 1.1.

1.3 *Scan* the labeled node $i \in V_2$. If i is exposed, go to Step 2. Otherwise, find the edge $(j, i) \in M$ and give node $j \in V_1$ the label i. Return to 1.1.

Step 2 (Augmentation). An augmenting path P has been found. Use the labels to backtrack from $j \in V_2$ to find the path.
Augment M. $M \leftarrow (M \cup P) \setminus (M \cap P)$. Remove all labels. Return to Step 1.

Step 3 (No Augmenting Path). Let V_1^+, V_2^+ be the nodes of V_1 and V_2 that are labeled and V_1^-, V_2^- the unlabeled nodes.

Theorem 4.3 *On termination of the algorithm,*
(i) $R = V_1^- \cup V_2^+$ is a node covering of the edges E of G.
(ii) $\mid M \mid = \mid R \mid$, and M is optimal.

Proof. (a) As no more nodes can be labeled, from Step 1.2 it follows that there is no edge from V_1^+ to V_2^-. This means that $V_1^- \cup V_2^+$ covers E.
(b) As no augmenting path is found, every node of V_2^+ is incident to an edge e of M, and from Step 1.3 the other endpoint is in V_1^+.
(c) Every node of V_1^- is incident to an edge e of M, as otherwise it would have received the label $*$ in Step 1.0. The other endpoint is necessarily in V_2^-, as otherwise the node would have been labeled in Step 1.2.
(d) Thus $\mid V_1^- \cup V_2^+ \mid \leq \mid M \mid$. But $\mid R \mid \geq \mid M \mid$ and thus $\mid R \mid = \mid M \mid$. ∎

Example 4.1 Consider the bipartite graph shown in Figure 4.2 and the initial matching $M = \{(3,8), (5,10)\}$. The algorithm leads to the labeling shown, and the construction of the set of alternating paths shown. Two alternating paths are found: $(1,8), (3,8), (3,7)$ and $(4,10), (5,10), (5,9)$.

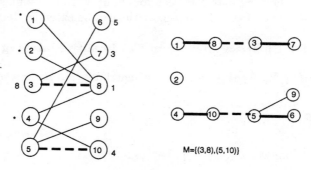

Fig. 4.2 Bipartite matching (Note: No odd Cycles)

In Figure 4.3 we show the new matching $M = \{(1,8), (3,7), (4,10), (5,9)\}$ and the labeling obtained from the algorithm. Now we see that we cannot add any more labels and no augmenting path has been found. It is easily checked that the node set $R = \{3, 4, 5, 8\}$ is an edge cover. As $\mid M \mid = \mid R \mid = 4$, M is optimal. ∎

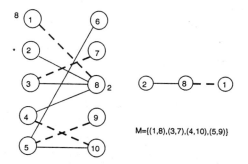

Fig. 4.3 Bipartite matching 2

4.3 THE ASSIGNMENT PROBLEM

Given a bipartite graph $G = (V_1, V_2, E)$ and weights c_e for $e \in E$, one problem is to find a matching of maximum weight. From the results on total unimodularity, we know that it suffices to solve the linear program:

$$z = \max \sum_{e \in E} c_e x_e$$
$$\sum_{e \in \delta(i)} x_e \leq 1 \text{ for } i \in V_1 \cup V_2$$
$$x_e \geq 0 \text{ for } e \in E.$$

When $|V_1| = |V_2| = n$, and the problem is to find a matching of size n of maximum weight, we obtain the assignment problem:

$$(P) \quad \begin{cases} z = \max \sum_{i=1}^n \sum_{j=1}^n c_{ij} x_{ij} \\ \sum_{j=1}^n x_{ij} = 1 \text{ for } i = 1, \ldots, n \\ \sum_{i=1}^n x_{ij} = 1 \text{ for } j = 1, \ldots, n \\ x_{ij} \geq 0 \text{ for } i, j = 1, \ldots, n. \end{cases}$$

Below we develop an algorithm for the assignment problem. Afterwards we show how the maximum weight bipartite matching problem can be solved as an assignment problem.

Taking the linear programming dual of the assignment problem, we get:

$$(D) \quad \begin{cases} w = \min \sum_{i=1}^n u_i + \sum_{j=1}^n v_j \\ u_i + v_j \geq c_{ij} \text{ for } i, j = 1, \ldots, n. \end{cases}$$

First we make an important observation that is valid for the assignment and traveling salesman problems allowing us to change the cost matrix in a certain way.

Proposition 4.4 *For all values $\{u_i\}_{i=1}^n$ and $\{v_j\}_{j=1}^n$, the value of any assignment with weights c_{ij} differs by a constant amount from its value with weights $\bar{c}_{ij} = c_{ij} - u_i - v_j$.*

58 MATCHINGS AND ASSIGNMENTS

This means that a solution is optimal with weights c_{ij} if and only if it is optimal with weights $\bar{c}_{ij}$.

Proof. For every primal feasible solution,

$$\sum_{i=1}^{n}\sum_{j=1}^{n}\bar{c}_{ij}x_{ij} = \sum_{i=1}^{n}\sum_{j=1}^{n}(c_{ij}-u_i-v_j)x_{ij}$$

$$= \sum_{i=1}^{n}\sum_{j=1}^{n}c_{ij}x_{ij} - \sum_{i=1}^{n}u_i(\sum_{j=1}^{n}x_{ij}) - \sum_{j=1}^{n}v_j(\sum_{i=1}^{n}x_{ij})$$

$$= \sum_{i=1}^{n}\sum_{j=1}^{n}c_{ij}x_{ij} - \sum_{i=1}^{n}u_i - \sum_{j=1}^{n}v_j.$$

So for any feasible solution of the primal, the difference in values of the two objectives are always the constant $\sum_{i=1}^{n}u_i + \sum_{j=1}^{n}v_j$, and the claim follows. ∎

Now we wish to characterize an optimal solution.

Proposition 4.5 *If there exist values u_i, v_j, and an assignment x_{ij} such that:*
(i) $\bar{c}_{ij} = c_{ij} - u_i - v_j \leq 0$, and
(ii) $x_{ij} = 1$ only when $\bar{c}_{ij} = 0$,
then the assignment x is optimal and has value $\sum_{i=1}^{n}u_i + \sum_{j=1}^{n}v_j$.

Proof. Because $\bar{c}_{ij} \leq 0$ for all i,j, the value of an optimal assignment with weights $\bar{c}_{ij}$ is necessarily nonpositive. But by condition (ii), with weights $\bar{c}_{ij}$, the assignment x has value $\sum_{i=1}^{n}\sum_{j=1}^{n}\bar{c}_{ij}x_{ij} = 0$ and is thus optimal. Now by Proposition 4.4, x is also optimal with weights c_{ij} and has value $\sum_{i=1}^{n}u_i + \sum_{j=1}^{n}v_j$. ∎

Note that this is another way of writing the linear programming complementarity conditions. In fact (i) tells us that u,v is a dual feasible solution, and (ii) that complementary slackness holds.

Idea of the Algorithm. We are going to use a so-called "primal-dual" algorithm on the graph $G = (V_1, V_2, E)$ where $V_1 = \{1, \cdots, n\}$, $V_2 = \{1', \cdots, n'\}$ and E consists of all edges with one endpoint in V_1 and the other in V_2.

At all times we will have a dual feasible solution u, v, or in other words $\bar{c}_{ij} \leq 0$ for all $i \in V_1, j \in V_2$.

Then we will try to find an assignment (a matching of size n) using only the edges $\overline{E} \subseteq E$ where $\overline{E} = \{(i,j) : \bar{c}_{ij} = 0\}$. To do this we will solve the maximum cardinality matching problem on the graph $\overline{G} = (V_1, V_2, \overline{E})$.

If we find a matching of size n, then by Proposition 4.5 we have an optimal weight assignment. Otherwise we return to the dual step and change the dual variables.

Algorithm for the Assignment Problem

Step 0. Let u, v be initial weights such that $\bar{c}_{ij} \leq 0$ for all i, j. Let $\overline{E} = \{(i,j) : \bar{c}_{ij} = 0\}$. Find a maximum cardinality matching M^* in the graph $G = (V_1, V_2, \overline{E})$ using the algorithm described in the previous section.
   If $\mid M^* \mid = n$, stop. M^* is optimal.
   Otherwise, note the matching $M = M^*$ and the labeled nodes V_1^+, V_2^+ on termination. Go to Step 2.

Step 1 (Primal Step). Let $\overline{E} = \{(i,j) : \bar{c}_{ij} = 0\}$. M is a feasible matching, and V_1^+, V_2^+ are feasible labels. Continue with the maximum cardinality matching algorithm of the previous section to find an optimal matching M^*.
   If $\mid M^* \mid = n$, stop. M^* is optimal.
   Otherwise, note the matching $M = M^*$ and the labeled nodes V_1^+, V_2^+ on termination. Go to Step 2.

Step 2 (Dual Step). Change the dual variables as follows:
   Set $\delta = \min_{i \in V_1^+, j \in V_2 \setminus V_2^+} [-\bar{c}_{ij}]$.
   Set $u_i \leftarrow u_i - \delta$ for $i \in V_1^+$.
   Set $v_j \leftarrow v_j + \delta$ for $j \in V_2^+$.
   Return to Step 1. ∎

We now need to show that the algorithm terminates correctly, and then see how long it takes.

Proposition 4.6 *Each time that Step 1 terminates with labeled nodes V_1^+, V_2^+, $\mid V_1^+ \mid > \mid V_2^+ \mid$.*

Proof. Every node of V_2^+ touches a matching edge whose other endpoint is in V_1^+. In addition V_1^+ contains at least one node that is not incident to the matching and received an initial label $*$. ∎

Proposition 4.7 *In the dual step, $\delta > 0$.*

Proof. We observed in the proof of Theorem 4.3 that there are no edges of $\overline{E}$ between V_1^+ and $V_2 \setminus V_2^+$. Therefore $\bar{c}_{ij} < 0$ for all $i \in V_1^+, j \in V_2 \setminus V_2^+$. ∎

Proposition 4.8 *After a dual change,*
$\bar{c}_{ij} \leftarrow \bar{c}_{ij}$ for $i \in V_1^+, j \in V_2^+$
$\bar{c}_{ij} \leftarrow \bar{c}_{ij}$ for $i \in V_1 \setminus V_1^+, j \in V_2 \setminus V_2^+$
$\bar{c}_{ij} \leftarrow \bar{c}_{ij} - \delta$ for $i \in V_1 \setminus V_1^+, j \in V_2^+$
$\bar{c}_{ij} \leftarrow \bar{c}_{ij} + \delta$ for $i \in V_1^+, j \in V_2 \setminus V_2^+$
and the new solution is dual feasible.

Proof. $\bar{c}_{ij}$ only increases when $i \in V_1^+, j \in V_2 \setminus V_2^+$. So we must check that the new values are nonpositive. However, δ was chosen precisely so that this condition is satisfied and at least one of these edges now has value $\bar{c}_{ij} = 0$. ∎

Proposition 4.9 *The labels V_1^+ and V_2^+ remain valid after a dual change.*

Proof. During a dual change, $\bar{c}_{ij}$ is unchanged for $i \in V_1^+, j \in V_2^+$, and so the labeling remains valid. ∎

The above observations tell us that the primal step can restart with the old labels, and the dual objective value decreases by a positive amount $\delta(|V_1^+| - |V_2^+|)$ at each iteration. However, to see that the algorithm runs fast, we can say much more.

Observation 4.2 $|V_2^+|$ increases after a dual change, because of the previous proposition and the choice of δ.

Observation 4.3 The cardinality of the maximum cardinality matching must increase after at most n dual changes, as $|V_2^+|$ cannot exceed n.

Observation 4.4 The cardinality of the maximum cardinality matching can increase at most n times, as $|M^*|$ cannot exceed n.

Proposition 4.10 *The algorithm has complexity $O(n^4)$.*

Proof. By the previous observations, the total number of dual changes in the course of the algorithm is $O(n^2)$. Work in a dual step is $O(|E|)$. The work in the primal step between two augmentations is also $O(|E|)$. ∎

Example 4.2 Consider an instance of the assignment problem with $n = 4$ and the profit matrix

$$(c_{ij}) = \begin{pmatrix} 27 & 17 & 7 & 8 \\ 14 & 2 & 10 & 2 \\ 12 & 19 & 4 & 4 \\ 8 & 6 & 12 & 6 \end{pmatrix}.$$

We apply the assignment algorithm. In step 0, we find a first dual feasible solution by setting $v_j^1 = \max_i c_{ij}$ for $j = 1, \ldots n$ and $u_i^1 = 0$ for $i = 1, \ldots, n$. This gives a dual feasible solution, and the reduced profit matrix

$$(\bar{c}_{ij}^1) = \begin{pmatrix} 0 & -2 & -5 & 0 \\ -13 & -17 & -2 & -6 \\ -15 & 0 & -8 & -4 \\ -19 & -13 & 0 & -2 \end{pmatrix} \text{ with } u = (0,0,0,0), v = (27,19,12,8).$$

The corresponding dual solution has value $\sum_{i=1}^{4} u_i^1 + \sum_{j=1}^{4} v_j^1 = 66$. Now we observe that there is no zero entry in the second row of this matrix, so we can immediately improve the dual solution, and more importantly add another edge to $\overline{E}$ by setting $u_2^2 = \max_j \bar{c}_{2j}^1 = -2$. This gives the new reduced profit matrix

$$(\bar{c}_{ij}^2) = \begin{pmatrix} 0 & -2 & -5 & 0 \\ -11 & -15 & 0 & -4 \\ -15 & 0 & -8 & -4 \\ -19 & -13 & 0 & -2 \end{pmatrix} \text{ with } u = (0, -2, 0, 0), v = (27, 19, 12, 8)$$

and a dual objective value of 64.

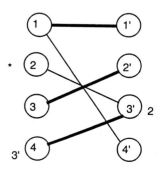

Fig. 4.4 Primal step

In the primal step we now construct the bipartite graph shown in Figure 4.4. We find an initial matching easily by a greedy approach. Suppose we obtain the matching $M = \{(1, 1'), (3, 2'), (4, 3')\}$ with $|M| = 3$. The augmenting path algorithm leads to the labels shown in Figure 4.4.

Node 2 receives the label *.
Node 3' receives the label 2.
Node 4 receives the label 3', and then no more labels can be given.
Thus $V_1^+ = \{2, 4\}$ and $V_2^+ = \{3'\}$.

In the dual step, we find that $\delta = -\bar{c}_{44}^2 = 2$. This leads to a modified dual feasible solution and the reduced profit matrix

$$(\bar{c}_{ij}^3) = \begin{pmatrix} 0 & -2 & -7 & 0 \\ -9 & -13 & 0 & -2 \\ -15 & 0 & -10 & -4 \\ -17 & -11 & 0 & 0 \end{pmatrix} \text{ with } u = (0, -4, 0, -2), v = (27, 19, 14, 8).$$

The dual solution now has value 62.

Now in the primal step the edge $(4, 4')$ is added to $\overline{E}$. The same labels can be given as before and in addition node 4' now receives the label 4. An augmenting path $\{(2, 3'), (4, 3'), (4, 4')\}$ has been found. $M = \{(1, 1'), (2, 3'), (3, 2'), (4, 4')\}$ is a larger matching, and as it is of size $n = 4$, it is optimal. It can also be checked that its value is 62, equal to that of the dual solution. ∎

62   MATCHINGS AND ASSIGNMENTS

Now we return to the maximum weight bipartite matching problem. We demonstrate by example how the assignment algorithm can be used to solve it. Consider the instance shown in Figure 4.5.

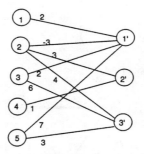

Fig. 4.5   Maximum weight bipartite matching

As $|V_1| - |V_2| = 2$, we add two nodes $4'$ and $5'$ to V_2. All existing edges are given a weight $\max\{c_e, 0\}$. All missing edges are added with a weight of 0. The resulting assignment weight matrix is

$$(c_{ij}) = \begin{pmatrix} 2 & 0 & 0 & 0 & 0 \\ 0 & 3 & 4 & 0 & 0 \\ 2 & 0 & 6 & 0 & 0 \\ 0 & 1 & 0 & 0 & 0 \\ 7 & 0 & 3 & 0 & 0 \end{pmatrix}.$$

The assignment algorithm terminates with an optimal solution $x_{15} = x_{22} = x_{33} = x_{44} = x_{51} = 1$. The edges with positive weight $(22), (33), (51)$ provide a solution of the matching problem.

4.4   NOTES

More detailed chapters on assignment problems are in [AhuMagOrl93], and on general matchings in [CooCunetal97] and [NemWol88]. The complete story on matchings is in the book [LovPlu86].

4.1 The results on alternating paths are from [Ber57] and [NorRab59].

4.3 The primal-dual algorithm can be found in [Kuh55], [ForFul62]. Primal dual algorithms for linear programming were also proposed in [DanForFul56]. Many of the first algorithms for combinatorial optimization problems were primal-dual, such as the matching algorithm [Edm65b] and the matroid intersection algorithm [Edm70]; see also [Law76] and [CooCunetal97].

4.5 EXERCISES

1. Find two augmenting paths for the matching M shown in the graph of Figure 4.6. Is the new matching M' obtained after augmentation optimal? Why?

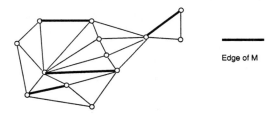

Fig. 4.6 Matching to be augmented

2. Find a maximum cardinality matching in the graph of Figure 4.7.

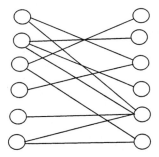

Fig. 4.7 Maximum cardinality matching

Demonstrate that your solution is optimal.

3. If a graph has $n = 2k$ nodes, a matching with k edges is called *perfect*. Show directly that the graph of Figure 4.7 does not contain a perfect matching.

4. Given a connected graph $G = (V, E)$ and positive edge lengths c_e for $e \in E$, the *Chinese Postman* (or *garbage collection*) *Problem* consists of visiting each edge of G at least once beginning and finishing at the same vertex, and minimizing the total distance traveled.

(i) Show that the minimum distance traveled is $\sum_{e \in E} c_e$ if and only if the graph is *Eulerian* (all nodes have even degree).
(ii) Show that if G is not Eulerian, and k is the number of nodes of odd degree, then k is even and at least $\frac{k}{2}$ edges must be traversed more than once.
(iii) Show that the minimum additional distance that must be traveled can be found by solving a minimum weight perfect matching problem in a certain

64 MATCHINGS AND ASSIGNMENTS

subgraph (suppose that if $e = (i, j) \in E$, the shortest path between i and j is via edge e).

5. Show how a maximum flow algorithm can be used to find a maximum cardinality matching in a bipartite graph.

6. Find a maximum weight assignment with the weight matrix:

$$(c_{ij}) = \begin{pmatrix} 6 & 2 & 3 & 4 & 1 \\ 9 & 2 & 7 & 6 & 0 \\ 8 & 2 & 1 & 4 & 9 \\ 2 & 1 & 3 & 4 & 4 \\ 1 & 6 & 2 & 9 & 1 \end{pmatrix}.$$

7. Find a maximum weight matching in the weighted bipartite graph of Figure 4.8.

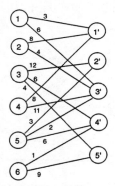

Fig. 4.8 Weighted bipartite graph

8. Show how an algorithm for the maximum weight bipartite matching problem can be used to solve the (equality) assignment problem.

9. Find a lower bound on the optimal value of a 6-city TSP instance with distance matrix

$$(c_{ij}) = \begin{pmatrix} - & 2 & 3 & 4 & 1 & 11 \\ 9 & - & 7 & 6 & 0 & 5 \\ 8 & 2 & - & 4 & 9 & 8 \\ 2 & 1 & 3 & - & 4 & 6 \\ 1 & 6 & 2 & 9 & - & 3 \\ 7 & 4 & 6 & 12 & 7 & - \end{pmatrix}.$$

10. Ten researchers are engaged in a set of ten projects. Let S_i denote the researchers working on project i for $i = 1, \ldots, 10$. To keep track of progress or problems, management wishes to designate one person working on each project to report at their weekly meeting. Ideally no person should be asked to report on more than one project. Is this possible or not, when $S_1 = \{3, 7, 8, 10\}, S_2 = \{4, 8\}, S_3 = \{2, 5, 7\}, S_4 = \{1, 2, 7, 9\}, S_5 = \{2, 5, 7\}, S_6 = \{1, 4, 5, 7\}, S_7 = \{2, 7\}, S_8 = \{1, 6, 7, 10\}, S_9 = \{2, 5\}, S_{10} = \{1, 2, 3, 6, 7, 8, 10\}$?

11. Suggest an algorithm to solve Exercise 5 of Section 1.9.

5
Dynamic Programming

5.1 SOME MOTIVATION: SHORTEST PATHS

Here we look at another approach to solving certain combinatorial optimization problems. To see the basic idea, consider the shortest path problem again. Given a directed graph $D = (V, A)$, nonnegative arc distances c_e for $e \in A$, and an initial node $s \in V$, the problem is to find the shortest path from s to every other node $v \in V \setminus \{s\}$. See Figure 5.1, in which a shortest path from s to t is shown, as well as one intermediate node p on the path.

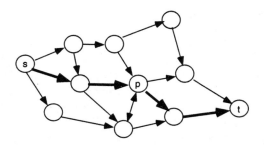

Fig. 5.1 Shortest s-t path

Observation 5.1 If the shortest path from s to t passes by node p, the subpaths (s, p) and (p, t) are shortest paths from s to p, and p to t respectively.

If this were not true, the shorter subpath would allow us to construct a shorter path from s to t, leading to a contradiction.

68 DYNAMIC PROGRAMMING

The question is how to use this idea to find shortest paths.

Observation 5.2 Let $d(v)$ denote the length of a shortest path from s to v. Then

$$d(v) = \min_{i \in V^-(v)} \{d(i) + c_{iv}\}. \tag{5.1}$$

In other words, if we know the lengths of the shortest paths from s to every neighbor (predecessor) of v, then we can find the length of the shortest path from s to v.

This still does not lead to an algorithm for general digraphs, because we may need to know $d(i)$ to calculate $d(j)$, and $d(j)$ to calculate $d(i)$. However, for certain digraphs, it does provide a simple algorithm.

Observation 5.3 Given an acyclic digraph $D = (V, A)$ with $n = |V|, m = |A|$, where the nodes are ordered so that $i < j$ for all arcs $(i, j) \in A$, then for the problem of finding shortest paths from node 1 to all other nodes, the recurrence (5.1) for $v = 2, \ldots, n$ leads to an $O(m)$ algorithm.

For arbitrary directed graphs with nonnegative weights $c \in R_+^{|A|}$, we need to somehow impose an ordering. One way to do this is to define a more general function $D_k(i)$ as the length of a shortest path from s to i containing at most k arcs. Then we have the recurrence:

$$D_k(j) = \min\{D_{k-1}(j), \min_{i \in V^-(j)} [D_{k-1}(i) + c_{ij}]\}.$$

Now by increasing k from 1 to $n-1$, and each time calculating $D_k(j)$ for all $j \in V$ by the recursion, we end up with an $O(mn)$ algorithm and $d(j) = D_{n-1}(j)$.

This approach whereby an optimal solution value for one problem is calculated recursively from the optimal values of slightly different problems is called *Dynamic Programming (DP)*. Below we will see how it is possible to apply similar ideas to derive a recursion for several interesting problems. The standard terminology used is the *Principle of Optimality* for the property that pieces of optimal solutions are themselves optimal, *states* that correspond to the nodes for which values need to be calculated, and *stages* for the steps which define the ordering.

5.2 UNCAPACITATED LOT-SIZING

The uncapacitated lot-sizing problem (ULS) was introduced in Chapter 1 where two different mixed integer programming formulations were presented. The problem again is to find a minimum cost production plan that satisfies

UNCAPACITATED LOT-SIZING 69

all the nonnegative demands $\{d_t\}_{t=1}^n$, given the costs of production $\{p_t\}_{t=1}^n$, storage $\{h_t\}_{t=1}^n$, and set-up $\{f_t\}_{t=1}^n$. We assume $f_t \geq 0$ for all t.

To obtain an efficient dynamic programming algorithm, it is necessary to understand the structure of the optimal solutions. For this it is useful to view the problem as a network design problem. Repeating the MIP formulation of Section 1.4, we have:

$$\min \sum_{t=1}^n p_t x_t + \sum_{t=1}^n h_t s_t + \sum_{t=1}^n f_t y_t \qquad (5.2)$$

$$s_{t-1} + x_t = d_t + s_t \text{ for } t = 1,\ldots,n \qquad (5.3)$$

$$x_t \leq M y_t \text{ for } t = 1,\ldots,n \qquad (5.4)$$

$$s_0 = s_n = 0, s \in R_+^{n+1}, x \in R_+^n, y \in B^n \qquad (5.5)$$

where x_t denotes the production in period t, and s_t the stock at the end of period t. We see that every feasible solution corresponds to a flow in the network shown in Figure 5.2,

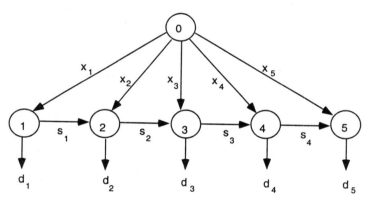

Fig. 5.2 Network for lot-sizing problem

where p_t is the flow cost on arc $(0,t)$, h_t is the flow cost on arc $(t,t+1)$, and the fixed costs f_t are incurred if $x_t > 0$ on arc $(0,t)$.

Thus ULS is a fixed charge network flow problem in which one must choose which arcs $(0,t)$ are open (which are the production periods), and then find a minimum cost flow through the network. Optimal solutions to ULS have two important structural properties that follow from this network viewpoint.

Proposition 5.1 (i) *There exists an optimal solution with $s_{t-1} x_t = 0$ for all t. (Production takes place only when the stock is zero.)*
(ii) *There exists an optimal solution such that if $x_t > 0$, $x_t = \sum_{i=t}^{t+k} d_i$ for some $k \geq 0$. (If production takes place in t, the amount produced exactly satisfies demand for periods t to $t+k$.)*

Proof. Suppose that the production periods have been chosen optimally (certain arcs $(0,t)$ are open). Now as an optimal extreme flow uses a set of arcs forming a tree, the set of arcs with positive flow contains no cycle, and it follows that only one of the arcs arriving at node t can have a positive flow (i.e., $s_{t-1}x_t = 0$). The second statement then follows immediately. ∎

Property (ii) is the important property we need to derive a DP algorithm for ULS.

For the rest of this section we let d_{it} denote the sum of demands for periods i up to t (i.e., $d_{it} = \sum_{j=i}^{t} d_j$). The next observation is repeated from Section 1.4 to simplify the calculations.

Observation 5.4 As $s_t = \sum_{i=1}^{t} x_i - d_{1t}$, the stock variables can be eliminated from the objective function giving $\sum_{t=1}^{n} p_t x_t + \sum_{t=1}^{n} h_t s_t = \sum_{t=1}^{n} p_t x_t + \sum_{t=1}^{n} h_t(\sum_{i=1}^{t} x_i - d_{1t}) = \sum_{t=1}^{n} c_t x_t - \sum_{t=1}^{n} h_t d_{1t}$ where $c_t = p_t + \sum_{i=t}^{n} h_i$. This allows us to work with the modified cost function $\sum_{t=1}^{n} c_t x_t + 0 \sum_{t=1}^{n} s_t + \sum_{t=1}^{n} f_t y_t$, and the constant term $\sum_{t=1}^{n} h_t d_{1t}$ must be subtracted at the end of the calculations.

Let $H(k)$ be the minimum cost of a solution for periods $1, \ldots, k$. If $t \leq k$ is the last period in which production occurs (namely $x_t = d_{tk}$), what happens in periods $1, \ldots, t-1$? Clearly the least cost solution must be optimal for periods $1, \ldots, t-1$, and thus has cost $H(t-1)$. This gives the recursion.

Forward Recursion

$$H(k) = \min_{1 \leq t \leq k} \{H(t-1) + f_t + c_t d_{tk}\}$$

with $H(0) = 0$.

Calculating $H(k)$ for $k = 1, \ldots, n$ leads to the value $H(n)$ of an optimal solution of ULS. Working back gives a corresponding optimal solution. It is also easy to see that $O(n^2)$ calculations suffice to obtain $H(n)$ and an optimal solution.

Example 5.1 Consider an instance of ULS with $n = 4, d = (2, 4, 5, 1), p = (3, 3, 3, 3), h = (1, 2, 1, 1)$ and $f = (12, 20, 16, 8)$. We start by calculating $c = (8, 7, 5, 4), (d_{11}, d_{12}, d_{13}, d_{14}) = (2, 6, 11, 12)$ and the constant $\sum_{t=1}^{4} h_t d_{1t} = 37$. Now we successively calculate the values of $H(k)$ using the recursion.
$H(1) = f_1 + c_1 d_1 = 28$.
$H(2) = \min[28 + c_1 d_2, H(1) + f_2 + c_2 d_2] = \min[60, 76] = 60$.
$H(3) = \min[60 + c_1 d_3, 76 + c_2 d_3, H(2) + f_3 + c_3 d_3] = \min[100, 111, 101] = 100$.
$H(4) = \min[100 + c_1 d_4, 111 + c_2 d_4, 101 + c_3 d_4, H(3) + f_4 + c_4 d_4]$
$\quad = \min[108, 118, 106, 112] = 106$.

Working backwards, we see that $H(4) = 106 = H(2) + f_3 + c_3 d_{34}$, so $y_3 = 1, x_3 = 6, y_4 = x_4 = 0$. Also $H(2) = f_1 + c_1 d_{12}$, so $y_1 = 1, x_1 = 6, y_2 = x_2 = 0$. Thus we have found an optimal solution $x = (6, 0, 6, 0), y = (1, 0, 1, 0), s =$

$(4, 0, 1, 0)$ whose value in the original costs is $106 - 37 = 69$. Checking we have $6p_1 + f_1 + 6p_3 + f_3 + 4h_1 + 1h_3 = 69$. ∎

Another possibility is to solve ULS directly as a shortest path problem. Consider a directed graph with nodes $\{0, 1, \ldots, n\}$ and arcs (i, j) for all $i < j$. The cost $f_{i+1} + c_{i+1}d_{i+1,j}$ of arc (i, j) is the cost of starting production in $i+1$ and satisfying the demand for periods $i+1$ up to j. Figure 5.3 shows the shortest path instance arising from the data of Example 5.1. Now a least cost path from node 0 to node n provides a minimum cost set of production intervals and solves ULS.

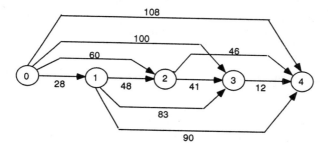

Fig. 5.3 Shortest path for ULS

We observe that $H(k)$ is the cost of a cheapest path from nodes 0 to k, and as the directed graph is acyclic, we know from Observation 5.3 that the corresponding shortest path algorithm is $O(m) = O(n^2)$.

5.3 AN OPTIMAL SUBTREE OF A TREE

Here we consider another problem that can be tackled by dynamic programming. However, the recursion here is not related at all to shortest paths. The *Optimal Subtree of a Tree Problem* involves a tree $T = (V, E)$ with a root $r \in V$ and weights c_v for $v \in V$. The problem is to choose a subtree of T rooted at r of maximum weight, or the empty tree if there is no positive weight rooted subtree.

To describe a dynamic programming recursion we need some notation. For a rooted tree, each node v has a well-defined *predecessor* $p(v)$ on the unique path from the root r to v, and, for $v \neq r$, a set of *immediate successors* $S(v) = \{w \in V : p(w) = v\}$. Also we let $T(v)$ be the *subtree of T rooted at v* containing all nodes w for which the path from r to w contains v.

For any node v of T, let $H(v)$ denote the optimal solution value of the rooted subtree problem defined on the tree $T(v)$ with node v as the root. If the optimal subtree is empty, clearly $H(v) = 0$. Otherwise the optimal subtree contains v. It may also contain subtrees of $T(w)$ rooted at w for $w \in S(v)$. By the principle of optimality, these subtrees must themselves be optimal rooted subtrees. Hence we obtain the recursion:

$$H(v) = \max\{0, c_v + \sum_{w \in S(v)} H(w)\}.$$

To initialize the recursion, we start with the leaves (nodes having no successors) of the tree. For a leaf $v \in V$, $H(v) = \max[c_v, 0]$. The calculations are then carried out by working in from the leaves to the root, until the optimal value $H(r)$ is obtained. As before, an optimal solution is then easily found by working backwards out from the root, eliminating every subtree $T(v)$ encountered with $H(v) = 0$. Finally note that each of the terms c_v and $H(v)$ occurs just once on the right-hand side during the recursive calculations, and so the algorithm is $O(n)$.

Example 5.2 For the instance of the optimal subtree of a tree problem shown in Figure 5.4 with root $r = 1$, we start with the leaf nodes $H(4) = H(6) = H(7) = H(11) = 0, H(9) = 5, H(10) = 3, H(12) = 3$, and $H(13) = 3$. Working in, $H(5) = \max[0, -6 + 5 + 3] = 2$ and $H(8) = \max[0, 2 + 0 + 3 + 3] = 8$. Now the values of $H(v)$ for all successors of nodes 2 and 3 are known, and so $H(2) = 4$ and $H(3) = 0$ can be calculated. Finally $H(1) = \max[0, -2+4+0] = 2$. Cutting off subtrees $T(3), T(4)$, and $T(6)$ leaves an optimal subtree with nodes 1,2,5,9,10 of value $H(1) = 2$. ∎

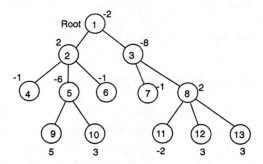

Fig. 5.4 Rooted tree with node weights c_v

5.4 KNAPSACK PROBLEMS

Here we examine various knapsack problems. Whereas ULS and the optimal subtree problems have the Efficient Optimization Property, knapsack problems in general are more difficult. This is made more precise in the next chapter. Dynamic programming provides an effective approach for such problems if the size of the data is restricted.

5.4.1 0–1 Knapsack

First we consider the 0–1 knapsack problem:

$$z = \max \sum_{j=1}^{n} c_j x_j$$
$$\sum_{j=1}^{n} a_j x_j \leq b$$
$$x \in B^n$$

where the coefficients $\{a_j\}_{j=1}^n$ and b are positive integers.

Thinking of the right-hand side λ taking values from $0, 1, \ldots, b$ as the state, and the subset of variables $x_1, \ldots, x_k$ represented by k as the stage, leads us to define the problem $P_r(\lambda)$ and the optimal value function $f_r(\lambda)$ as follows:

$(P_r(\lambda))$
$$f_r(\lambda) = \max \sum_{j=1}^{r} c_j x_j$$
$$\sum_{j=1}^{r} a_j x_j \leq \lambda$$
$$x \in B^r.$$

Then $z = f_n(b)$ gives us the optimal value of the knapsack problem. Thus we need to define a recursion that allows us to calculate $f_r(\lambda)$ in terms of values of $f_s(\mu)$ for $s \leq r$ and $\mu < \lambda$.

What can we say about an optimal solution x^* for problem $P_r(\lambda)$ with value $f_r(\lambda)$? Clearly either $x_r^* = 0$ or $x_r^* = 1$.

(i) If $x_r^* = 0$, then by the same optimality argument we used for shortest paths, $f_r(\lambda) = f_{r-1}(\lambda)$.

(ii) If $x_r^* = 1$, then $f_r(\lambda) = c_r + f_{r-1}(\lambda - a_r)$.

Thus we arrive at the recursion:

$$f_r(\lambda) = \max\{f_{r-1}(\lambda), c_r + f_{r-1}(\lambda - a_r)\}.$$

Now starting the recursion with $f_0(\lambda) = 0$ for $\lambda \geq 0$, or alternatively with $f_1(\lambda) = 0$ for $0 \leq \lambda < a_1$ and $f_1(\lambda) = \max[c_1, 0]$ for $\lambda \geq a_1$, we then use the recursion to successively calculate $f_2, f_3, \ldots, f_n$ for all integral values of λ from 0 to b.

The question that then remains is how to find an associated optimal solution. For this we have two related options. In both cases we iterate back from the optimal value $f_n(b)$. Either we must keep all the $f_r(\lambda)$ values, or an indicator $p_r(\lambda)$ which is 0 if $f_r(\lambda) = f_{r-1}(\lambda)$, and 1 otherwise.

If $p_n(b) = 0$, then as $f_n(b) = f_{n-1}(b)$, we set $x_n^* = 0$ and continue by looking for an optimal solution of value $f_{n-1}(b)$.

If $p_n(b) = 1$, then as $f_n(b) = c_n + f_{n-1}(b - a_n)$, we set $x_n^* = 1$ and then look for an optimal solution of value $f_{n-1}(b - a_n)$.

Iterating n times allows us to obtain an optimal solution.

74   DYNAMIC PROGRAMMING

Counting the number of calculations required to arrive at $z = f_n(b)$, we see that for each calculation $f_r(\lambda)$ for $\lambda = 0, 1, \ldots, b$ and $r = 1, \ldots, n$ there are a constant number of additions, subtractions, and comparisons. Calculating the optimal solution requires at most the same amount of work. Thus the DP algorithm is $O(nb)$.

Example 5.3 Consider the 0–1 knapsack instance:

$$z = \max 10x_1 + 7x_2 + 25x_3 + 24x_4$$
$$2x_1 + 1x_2 + 6x_3 + 5x_4 \leq 7$$
$$x \in B^4.$$

The values of $f_r(\lambda)$ and $p_r(\lambda)$ are shown in Table 5.1. The values of $f_1(\lambda)$ are calculated by the formula described above. The next column is then calculated from top to bottom using the recursion. For example, $f_2(7) = \max\{f_1(7), 7 + f_1(7-1)\} = \max\{10, 7+10\} = 17$, and as the second term of the maximization gives the value of $f_2(7)$, we set $p_2(7) = 1$. The optimal value $z = f_4(7) = 34$.

	f_1	f_2	f_3	f_4	p_1	p_2	p_3	p_4
$\lambda = 0$	0	0	0	0	0	0	0	0
1	0	7	7	7	0	1	0	0
2	10	10	10	10	1	0	0	0
3	10	17	17	17	1	1	0	0
4	10	17	17	17	1	1	0	0
5	10	17	17	24	1	1	0	1
6	10	17	25	31	1	1	1	1
7	10	17	32	34	1	1	1	1

Table 5.1 $f_r(\lambda)$ for a 0–1 knapsack problem

Working backwards, $p_4(7) = 1$ and hence $x_4^* = 1$. $p_3(7-5) = p_3(2) = p_2(2) = 0$ and hence $x_3^* = x_2^* = 0$. $p_1(2) = 1$ and hence $x_1^* = 1$. Thus $x^* = (1, 0, 0, 1)$ is an optimal solution. ∎

5.4.2 Integer Knapsack Problems

Now we consider the integer knapsack problem:

$$z = \max \sum_{j=1}^{n} c_j x_j$$
$$\sum_{j=1}^{n} a_j x_j \leq b$$
$$x \in Z_+^n$$

where again the coefficients $\{a_j\}_{j=1}^n$ and b are positive integers. Copying from the 0–1 case, we define $P_r(\lambda)$ and the value function $g_r(\lambda)$ as follows:

$(P_r(\lambda))$
$$g_r(\lambda) = \max \sum_{j=1}^r c_j x_j \\ \sum_{j=1}^r a_j x_j \leq \lambda \\ x \in Z_+^r.$$

Then $z = g_n(b)$ gives us the optimal value of the integer knapsack problem.

To build a recursion, a first idea is to again copy from the 0–1 case. If x^* is an optimal solution to $P_r(\lambda)$ giving value $g_r(\lambda)$, then we consider the value of x_r^*. If $x_r^* = t$, then using the principle of optimality $g_r(\lambda) = c_r t + g_{r-1}(\lambda - t a_r)$ for some $t = 0, 1, \ldots, \lfloor \frac{\lambda}{a_r} \rfloor$, and we obtain the recursion:

$$g_r(\lambda) = \max_{t=0,1,\ldots,\lfloor \lambda/a_r \rfloor} \{c_r t + g_{r-1}(\lambda - t a_r)\}.$$

As $\lfloor \frac{\lambda}{a_r} \rfloor = b$ in the worst case, this gives an algorithm of complexity $O(nb^2)$.

Can one do better? Is it possible to reduce the calculation of $g_r(\lambda)$ to a comparison of only two cases?

(i) Taking $x_r^* = 0$, we again have $g_r(\lambda) = g_{r-1}(\lambda)$.

(ii) Otherwise, we must have $x_r^* \geq 1$, and we can no longer copy from above.

However, as $x_r^* = 1 + t$ with t a nonnegative integer, we claim, again by the principle of optimality, that if we reduce the value of x_r^* by 1, the remaining vector $(x_1^*, \ldots, x_{r-1}^*, t)$ must be optimal for the problem $P_r(\lambda - a_r)$. Thus we have $g_r(\lambda) = c_r + g_r(\lambda - a_r)$, and we arrive at the recursion:

$$g_r(\lambda) = \max\{g_{r-1}(\lambda), c_r + g_r(\lambda - a_r)\}.$$

This now gives an algorithm of complexity $O(nb)$, which is the same as that of the 0–1 problem. We again set $p_r(\lambda) = 0$ if $g_r(\lambda) = g_{r-1}(\lambda)$ and $p_r(\lambda) = 1$ otherwise.

Example 5.4 Consider the knapsack instance:

$$z = \max 7x_1 + 9x_2 + 2x_3 + 15x_4 \\ 3x_1 + 4x_2 + 1x_3 + 7x_4 \leq 10 \\ x \in Z_+^4.$$

The values of $g_r(\lambda)$ are shown in Table 5.2. With $c_1 \geq 0$, the values of $g_1(\lambda)$ are easily calculated to be $c_1 \lfloor \frac{\lambda}{a_1} \rfloor$. The next column is then calculated from top to bottom using the recursion. For example, $g_2(8) = \max\{g_1(8), 9 + g_2(8-4)\} = \max\{14, 9+9\} = 18$.

Working back, we see that $p_4(10) = p_3(10) = 0$ and thus $x_4^* = x_3^* = 0$. $p_2(10) = 1$ and $p_2(6) = 0$ and so $x_2^* = 1$. $p_1(6) = p_1(3) = 1$ and thus $x_1^* = 2$. Hence we obtain an optimal solution $x^* = (2, 1, 0, 0)$. ∎

76 DYNAMIC PROGRAMMING

	g_1	g_2	g_3	g_4 ‖ p_1	p_2	p_3	p_4
$\lambda = 0$	0	0	0	0 ‖ 0	0	0	0
1	0	0	2	2 ‖ 0	0	1	0
2	0	0	4	4 ‖ 0	0	1	0
3	7	7	7	7 ‖ 1	0	0	0
4	7	9	9	9 ‖ 1	1	0	0
5	7	9	11	11 ‖ 1	1	1	0
6	14	14	14	14 ‖ 1	0	0	0
7	14	16	18	18 ‖ 1	1	1	0
8	14	18	18	18 ‖ 1	1	0	0
9	21	21	21	21 ‖ 1	0	0	0
10	21	23	23	23 ‖ 1	1	0	0

Table 5.2 $g_r(\lambda)$ for an integer knapsack problem

Another recursion can be used for the integer knapsack problem. Looking at the example above, we see that in fact all the important information is contained in the nth column containing the values of $g_n(\lambda)$. Writing h in place of g_n, can we directly write a recursion for $h(\lambda)$?

Again the principle of optimality tells us that if x^* is an optimal solution of $P_n(\lambda)$ of value

$$h(\lambda) = \max\{\sum_{j=1}^{n} c_j x_j : \sum_{j=1}^{n} a_j x_j \leq \lambda, x \in Z_+^n\}$$

with $x_j^* \geq 1$, then $h(\lambda) = c_j + h(\lambda - a_j)$.

Thus we obtain the recursion:

$$h(\lambda) = \max[0, \max_{j:a_j \leq \lambda} \{c_j + h(\lambda - a_j)\}].$$

This also leads to an $O(nb)$ algorithm. Applied to the instance of Example 5.4, it gives precisely the values in the g_4 column of Table 5.2.

As a final observation, the dynamic programming approach for knapsack problems can also be viewed as a longest path problem. Construct an acyclic digraph $D = (V, A)$ with nodes $0, 1, \ldots, b$, arcs $(\lambda, \lambda + a_j)$ for $\lambda \in Z_+^1, \lambda \leq b - a_j$ with weight c_j for $j = 1, \ldots, n$, and 0-weight arcs $(\lambda, \lambda + 1)$ for $\lambda \in Z_+^1, \lambda \leq b - 1$. $h(\lambda)$ is precisely the value of a longest path from node 0 to node λ. Figure 5.5 shows the digraph arising from the instance:

$$z = \max 10x_1 + 7x_2 + 25x_3 + 24x_4$$
$$2x_1 + 1x_2 + 6x_3 + 5x_4 \leq 7$$
$$x \in Z_+^4,$$

except that the 0-weight arcs $(\lambda, \lambda + 1)$ are omitted by dominance.

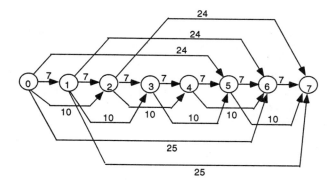

Fig. 5.5 Knapsack longest path problem

5.5 NOTES

5.1 The principle of optimality and dynamic programming originated with Bellman [Bell57]. One of the more recent books on dynamic programming is [Den82]. Shortest path problems with additional restrictions arise as subproblems in many routing problems, and are solved by dynamic programming, see [Desetal95].

5.2 The uncapacitated lot-sizing model and the dynamic programming algorithm for it are from [WagWhi58]. Dynamic programming recursions have been developed for many generalizations including lot-sizing with backlogging [Zan66], with constant production capacities [FloKle71], with start-up costs [Fle90], and with lower bounds on production [Con98]. Recently [WagvanHKoe92] among others have shown how the running time of the DP algorithm for the basic model can be significantly improved.

5.3 Various generalizations of the subtree problem are of interest. The more general problem of finding an optimal upper/lower set in a partially ordered set is examined in [GroLie81], and shown to be solvable as a network flow problem.

The problem of partitioning a tree into subtrees is tackled using dynamic programming in [BarEdmWol86], and this model is further studied in [AghMagWol95]. A telecommunications problem with such structure is studied in [Balaketal95].

Many other combinatorial optimization problems become easy when the underlying graph is a tree [MagWol95], or more generally a series-parallel graph [Taketal82].

5.4 The classical paper on the solution of knapsack problems by dynamic programming is [GilGom66]. The longest path or dynamic programming viewpoint was later extended by Gomory leading to the group relaxation of an

78   DYNAMIC PROGRAMMING

integer program [Gom65], and later to the superadditive duality theory for integer programs; see Notes of Section 2.2. The idea of reversing the roles of the objective and constraint rows (Exercise 5.7) is from [IbaKim75].

The dynamic programming appoach to TSP was first described by [HelKar62]. Relaxations of this approach, known as *state space relaxation*, have been used for a variety of constrained path and routing problems; see [ChrMinTot81]. Recently [Psa80] and [Balas95] have shown how such a recursion is of practical interest when certain restrictions on the tours or arrival sequences are imposed.

5.6   EXERCISES

1. Solve the uncapacitated lot-sizing problem with $n = 4$ periods, unit production costs $p = (1, 1, 1, 2)$, unit storage costs $h = (1, 1, 1, 1)$, set-up costs $f = (20, 10, 45, 15)$, and demands $d = (8, 5, 13, 4)$.

2. Consider the uncapacitated lot-sizing problem with backlogging ($ULSB$). *Backlogging* means that demand in a given period can be satisfied from production in a later period. If $r_t \geq 0$ denotes the amount backlogged in period t, the flow conservation constraints (5.3) become

$$s_{t-1} - r_{t-1} + x_t = d_t + s_t - r_t.$$

Show that there always exists an optimal solution to $ULSB$ with

(i) $s_{t-1} x_t = x_t r_t = s_{t-1} r_t = 0$.
(ii) $x_t > 0$ implies $x_t = \sum_{i=p}^{q} d_i$ with $p \leq t \leq q$.

Use this to derive a dynamic programming recursion for $ULSB$, or to reformulate $ULSB$ as a shortest path problem.

3. Find a maximum weight rooted subtree for the rooted tree shown in Figure 5.6.

4. Formulate the optimal subtree of a tree problem as an integer program. Is this IP easy to solve?

5. Given a digraph $D = (V, A)$, travel times c_{ij} for $(i, j) \in A$ for traversing the arcs, and earliest passage times r_j for $j \in V$, consider the problem of minimizing the time required to go from node 1 to node n.

(i) Describe a dynamic programming solution.
(ii) Formulate as a mixed integer program. Is this mixed integer program easy to solve?

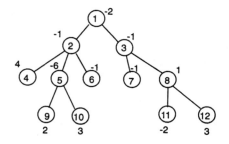

Fig. 5.6 Rooted tree with node weights c_v

6. Solve the knapsack problem

$$\min 5x_1 - 3x_2 + 7x_3 + 10x_4$$
$$2x_1 - x_2 + 4x_3 + 5x_4 \geq \lambda$$
$$x_1, x_4 \in Z_+^1, x_2, x_3 \in B^1$$

for all values of λ between 7 and 10.

7. Let $f(\lambda) = \max\{\sum_{j=1}^n c_j x_j : \sum_{j=1}^n a_j x_j \leq \lambda, x \in X \subset R^n\}$ and $h(t) = \min\{\sum_{j=1}^n a_j x_j : \sum_{j=1}^n c_j x_j \geq t, x \in X\}$.
Show that

(i) $f(\lambda) \geq t$ if and only if $h(t) \leq \lambda$.
(ii) $f(b) = \max\{t : h(t) \leq b\}$.

Use these observations to solve the knapsack problem

$$\max 3x_1 + 4x_2 + 6x_3 + 5x_4 + 8x_5$$
$$412x_1 + 507x_2 + 714x_3 + 671x_4 + 920x_5 \leq 1794$$
$$x \in B^5$$

by dynamic programming.

8. Solve the problem

$$\max x_1^3 + 2x_2^{\frac{1}{2}} + 4x_3 + 4x_4$$
$$2x_1^2 + 4x_2 + 6x_3 + 5x_4 \leq t$$
$$x_1 \leq 3, x_2 \leq 2, x_4 \leq 1$$
$$x \in Z_+^4$$

for $t \leq 12$. (Hint. One of the recursions in the chapter is appropriate.)

9. Formulate the 0–1 knapsack problem as a longest path problem.

10. Derive a dynamic programming recursion for the $STSP$ using $f(S,j)$ with $1,j \in S$ where $f(S,j)$ denotes the length of a shortest Hamiltonian path starting at node 1, passing through the nodes of $S \setminus \{1,j\}$ and terminating at node j.

11. Given the weighted rooted subtree of Figure 5.6, devise a dynamic programming algorithm to find optimal weighted subtrees with $k = 2, 3, \ldots, n-1$ nodes.

12. Given a tree T and a list of $T_1, \ldots, T_m$ of subtrees of T with weights $c(T_i)$ for $i = 1, \ldots, m$, describe an algorithm to find a maximum weight packing of node disjoint subtrees.

13.* Given a rooted tree T on n nodes, and an n by n matrix C, describe an algorithm to find a maximum weight packing of rooted subtrees of T, where the value of a subtree on node set S with root $i \in S$ is $\sum_{j \in S} c_{ij}$.

6
Complexity and Problem Reductions

6.1 COMPLEXITY

If we consider a list of the problems we have examined so far, we have either shown or it can be shown that the following have the Efficient Optimization Property:

The Uncapacitated Lot-Sizing Problem (Chapter 5)
The Maximum Weight Tree Problem (Chapter 3)
The Maximum Weight Matching Problem
The Shortest Path Problem (Chapter 5)
The Max Flow Problem (Chapter 3)
The TU Integer Programming Problem (Chapter 3)
The Assignment Problem (Chapter 4)

Below we make this more precise: there is a polynomial algorithm for these optimization problems.

On the other hand, no one to date has found an efficient (polynomial) algorithm for any of the following optimization problems:

The 0–1 Knapsack Problem (Chapter 1)
The Set Covering Problem (Chapter 1)
The Traveling Salesman Problem (Chapter 1)
The Uncapacitated Facility Location Problem (Chapter 1)
The Integer Programming Problem (Chapter 1)
The Steiner Tree Problem (Chapter 3)

Capacitated Lot Sizing (CLS) p.86

The remainder of this book will in large part be devoted to examining how to tackle problems in this second group. However, it is first useful to discuss the distinction (real or imaginary) between these two groups, so that when we encounter a new optimization problem we have an idea how we might classify and then attempt to solve it.

To develop such a method of classification, we need just four concepts:

A class $\mathcal{C}$ of legitimate problems to which the theory applies
A nonempty subclass $\mathcal{C}_A \subseteq \mathcal{C}$ of "easy" problems
A nonempty subclass $\mathcal{C}_B \subseteq \mathcal{C}$ of "difficult" problems
A relation "not more difficult than" between pairs of legitimate problems.

This immediately leads to:

Proposition 6.1 *(Reduction Lemma) Suppose that P and Q are two legitimate problems.*
If Q is "easy" and P is "not more difficult than" Q, then P is "easy".
If P is "difficult" and P is "not more difficult than" Q, then Q is "difficult".

We have already used the first part of the lemma implicitly in Chapter 4. There we show that the maximum weight bipartite matching problem is "easy" by showing that it is reducible to the assignment problem. Also Exercise 4.4 involves showing that the Chinese postman problem is reducible to maximum weight matching. The goal of the rest of this chapter is to somewhat formalize these notions. In the next section we introduce the class of legitimate problems and the "easy" class, and in Section 6.3 we discuss the concept of problem reduction which allows us to then define the "difficult" class. By the end of the chapter we will then have another tool at our disposal: namely a way to show that certain problems are "difficult" by using the second part of the reduction lemma.

6.2 DECISION PROBLEMS, AND CLASSES $\mathcal{NP}$ AND $\mathcal{P}$

Unfortunately, the theory does not exactly address optimization problems in the form we have posed them so far. To define the class of legitimate problems, it is appropriate to pose decison problems having YES-NO answers. Thus an optimization problem:
$$\max\{cx : x \in S\}$$
for which an instance consists of: $\{c$ and a "standard" representation of $S\}$ is replaced by the *decision problem*:

Is there an $x \in S$ with value $cx \geq k$?

for which an instance consists of $\{c$, a "standard" representation of S, and an integer $k\}$.

DECISION PROBLEMS, AND CLASSES $\mathcal{NP}$ AND $\mathcal{P}$

So for the rest of this chapter (unless explicitly stated), when we refer to an optimization problem TSP, UFL, and so on, we have in mind the corresponding decision problem.

Next we give a slightly more formal definition of the running time of an algorithm than that given at the start of Chapter 3. It is not just the number of variables and constraints or nodes and edges that defines the input length, but also the size of the numbers occurring in the data.

Definition 6.1 For a problem instance X, the *length of the input* $L = L(X)$ is the length of the binary representation of a "standard" representation of the instance.

Definition 6.2 Given a problem P, an algorithm A for the problem, and an instance X, let $f_A(X)$ be the number of elementary calculations required to run the algorithm A on the instance X. $f_A^*(l) = \sup_X\{f_A(X) : |X| = l\}$ is the *running time* of algorithm A. An algorithm A is *polynomial* for a problem P if $f_A^*(l) = O(l^p)$ for some positive integer p.

Now we can define the class of "legitimate" problems.

Definition 6.3 $\mathcal{NP}$ is the class of decision problems with the property that: for any instance for which the answer is YES, there is a "short" (polynomial) proof of the YES.

We note immediately that if a decision problem associated to an optimization problem is in $\mathcal{NP}$, then the optimization problem can be solved by answering the decision problem a polynomial number of times (by using bisection on the objective function value).

Proposition 6.2 For each optimization problem in the two lists in Section 6.1, the associated decision problem: "Does there exist a primal solution of value as good as or better than k?" lies in $\mathcal{NP}$.

Now we can define the class of "easy" problems.

Definition 6.4 $\mathcal{P}$ is the class of decision problems in $\mathcal{NP}$ for which there exists a polynomial algorithm.

Example 6.1 (i) Uncapacitated Lot Sizing. Consider ULS for which a dynamic programming algorithm is presented in Chapter 5. For an instance X with integral data (n, d, p, h, f, k), the input has length $L(X) = \sum_{j=1}^{n} \lceil \log d_j \rceil + \sum_{j=1}^{n} \lceil \log p_j \rceil + \sum_{j=1}^{n} \lceil \log h_j \rceil + \sum_{j=1}^{n} \lceil \log f_j \rceil + \lceil \log k \rceil$.

The DP algorithm requires only $O(n^2)$ additions and comparisons of the numbers occurring in the data, and hence the size of numbers required to give a YES answer, and the running time are certainly $O(L^2)$. Thus ULS is in $\mathcal{P}$.

(ii) 0–1 Knapsack. For an instance X of 0–1 *KNAPSACK*: $\{\sum_{j=1}^{n} c_j x_j \geq k, \sum_{j=1}^{n} a_j x_j \leq b, x \in \{0,1\}^n\}$, the length of the input is $L(X) = \sum_{j=1}^{n} \lceil \log c_j \rceil + \sum_{j=1}^{n} \lceil \log a_j \rceil + \lceil \log b \rceil + \lceil \log k \rceil$.

For an instance for which the answer is YES, it suffices to (a) read a solution $x^* \in \{0,1\}^n$, and (b) check that $ax^* \leq b$ and $cx^* \geq k$. Both (a) and (b) can be carried out in time polynomial in L, so the associated decision problem is in $\mathcal{NP}$.

From Section 5.4, dynamic programming provides an $O(nb)$ algorithm. As b is not equal to $(\log b)^p$ for any fixed p, this algorithm is not polynomial, and in fact no polynomial algorithm is known for 0–1 *KNAPSACK*.

(iii) Symmetric Traveling Salesman. For *STSP* with instance $(G, c^{|E|}, k)$, it suffices to check that a proposed set of edges forms a tour and that its length does not exceed k. The argument for most other problems on the second list is similar.

(iv) Integer Programming. Problem *IP* requires a little more work, because one needs to show that there always exists an optimal solution x^* whose description length $\sum_{j=1}^{n} \lceil \log x_j^* \rceil$ is polynomial in L. ∎

Do the second set of optimization problems listed in Section 6.1 above have anything in common apart from the fact that their decision problems are in $\mathcal{NP}$, and that nobody has yet discovered a polynomial algorithm for any of them?

Surprisingly, they have a second property in common: their decision problems are all among the *most difficult problems* in $\mathcal{NP}$.

6.3 POLYNOMIAL REDUCTION AND THE CLASS $\mathcal{NPC}$

This is the formal definition of "is not more difficult than" that we need.

Definition 6.5 If $P, Q \in \mathcal{NP}$, and if an instance of P can be converted in polynomial time to an instance of Q, then P is *polynomially reducible* to Q.

Note that this means that if we have an algorithm for problem Q, it can be used to solve problem P with an overhead that is polynomial in the size of the instance. We now define the class of "most difficult" problems.

Definition 6.6 $\mathcal{NPC}$, the class of $\mathcal{NP}$-*complete* problems, is the subset of problems $P \in \mathcal{NP}$ such that for all $Q \in \mathcal{NP}$, Q is polynomially reducible to P.

It is a remarkable fact not only that $\mathcal{NPC}$ is nonempty, but that all of the decision problems in our second list are in $\mathcal{NPC}$. So how can one prove that a problem is in $\mathcal{NPC}$?

The most important step is to prove that $\mathcal{NPC}$ is nonempty. Written as an 0–1 integer program, $SATISFIABILITY$ is the decision problem:

Given $N = \{1, \ldots, n\}$, and $2m$ subsets $\{C_i\}_{i=1}^{m}$ and $\{D_i\}_{i=1}^{m}$ of N, does the 0–1 integer program:

$$\sum_{j \in C_i} x_j + \sum_{j \in D_i} (1 - x_j) \geq 1 \text{ for } i = 1, \ldots, m$$

$$x \in B^n$$

have a feasible solution?

It is obvious that this problem is in $\mathcal{NP}$. Cook showed in 1970 that $SATISFIABILITY$ is in $\mathcal{NPC}$.

Now we indicate how the reduction lemma can be used to show that all the problems of the second list and many others are in $\mathcal{NPC}$.

For example, to see that BIP is in $\mathcal{NPC}$, all we need to observe is that

(i) $BIP \in \mathcal{NP}$, (which is immediate) and

(ii) $SATISFIABILITY$ reduces to BIP. (Above we actually described $SATISFIABILITY$ as a BIP, and so this is also immediate).

We now restate the Reduction Lemma (Proposition 6.1) more formally.

Proposition 6.3 *Suppose that problems $P, Q \in \mathcal{NP}$.*
(i) *If $Q \in \mathcal{P}$ and P is polynomially reducible to Q, then $P \in \mathcal{P}$.*
(ii) *If $P \in \mathcal{NPC}$ and P is polynomially reducible to Q, then $Q \in \mathcal{NPC}$.*

Proof. (ii) Consider any problem $R \in \mathcal{NP}$. As $P \in \mathcal{NPC}$, R is polynomially reducible to P. However P is polynomially reducible to Q by hypothesis, and thus R is polynomially reducible to Q. As this holds for all $R \in \mathcal{NP}$, $Q \in \mathcal{NPC}$. ∎

This has an important corollary.

Corollary 6.1 *If $\mathcal{P} \cap \mathcal{NPC} \neq \emptyset$, then $\mathcal{P} = \mathcal{NP}$.*

Proof. Suppose $Q \in \mathcal{P} \cap \mathcal{NPC}$ and take $R \in \mathcal{NP}$. By (ii), as $R \in \mathcal{NP}$ and $Q \in \mathcal{NPC}$, R is polynomially reducible to Q. By (i), as $Q \in \mathcal{NP}$ and R is polynomially reducible to Q, $R \in \mathcal{P}$. So $\mathcal{NP} \subseteq \mathcal{P}$ and thus $\mathcal{P} = \mathcal{NP}$. ∎

The list of problems known to be in $\mathcal{NPC}$ is now enormous. Some of the most basic problems in $\mathcal{NPC}$ are the problems in our second list. Below we prove that the *Capacitated Lot-Sizing Problem (CLS)* can be added to the list.

Example 6.2 Consider the lot-sizing problem introduced in Chapter 1, with a production capacity constraint in each period. CLS has a formulation:

$$\min \sum_{t=1}^n p_t x_t + \sum_{t=1}^n h_t s_t + \sum_{t=1}^n f_t y_t$$
$$s_{t-1} + x_t = d_t + s_t \text{ for } t = 1, \ldots, n$$
$$x_t \leq C_t y_t \text{ for } t = 1, \ldots, n$$
$$s_0 = s_n = 0, s \in R_+^{n+1}, x \in R_+^n, y \in B^n.$$

First, is the decision version of CLS in $\mathcal{NP}$? One answer lies in the observation that there is always an optimal solution of the form: $y \in \{0,1\}^n$ with x a basic feasible solution of the network flow problem in which y is fixed. So there exists an optimal solution whose length is polynomial in the length of the input, and can be used to verify a YES answer in polynomial time. One just needs to check that it satisfies the constraints and its value is sufficiently small.

We now show that 0–1 $KNAPSACK$ is polynomially reducible to CLS. To do this, we show how an instance of the 0–1 knapsack problem can be solved as a capacitated lot-sizing problem. Given an instance:

$$\min\{\sum_{j=1}^n c_j y_j : \sum_{j=1}^n a_j y_j \geq b, y \in B^n\},$$

we solve a lot-sizing instance with n periods, $p_t = h_t = 0, f_t = c_t, C_t = a_t$ for all t, $d_t = 0$ for $t = 1, \ldots, n-1$ and $d_n = b$.

An equivalent formulation of the lot-sizing problem, obtained by eliminating the stock variables as described in Section 1.4, is:

$$\min \sum_{t=1}^n p'_t x_t + \sum_{t=1}^n f_t y_t$$
$$\sum_{i=1}^t x_i \geq \sum_{i=1}^t d_i \text{ for } t = 1, \ldots, n-1$$
$$\sum_{i=1}^n x_i = \sum_{i=1}^n d_i$$
$$x_t \leq C_t y_t \text{ for } t = 1, \ldots, n$$
$$x \in R_+^n, y \in B^n.$$

Rewriting this with the chosen values for the data, we obtain:

$$\min \sum_{t=1}^n c_t y_t$$
$$\sum_{i=1}^t x_i \geq 0 \text{ for } t = 1, \ldots, n-1$$

$$\sum_{i=1}^{n} x_i = b$$
$$x_t \leq a_t y_t \text{ for } t = 1, \ldots, n$$
$$x \in R_+^n, y \in B^n.$$

Dropping the $n-1$ redundant demand constraints leaves

$$\min \sum_{t=1}^{n} c_t y_t$$
$$\sum_{t=1}^{n} x_t = b$$
$$x_t \leq a_t y_t \text{ for } t = 1, \ldots, n$$
$$x \in R_+^n, y \in B^n.$$

Now let (x^*, y^*) be an optimal solution of this lot-sizing instance. Combining the constraint $\sum_{t=1}^{n} x_t = b$ with the constraints $x_t \leq a_t y_t$ for $t = 1, \ldots, n$, we see that $\sum_t a_t y_t^* \geq b$ and so y^* is feasible in the knapsack instance. It is also optimal, because a better knapsack solution $\overline{y}$ with $c\overline{y} < cy^*$ would also provide a better lot-sizing solution $((\overline{x})_t = a_t \overline{y}_t, \overline{y})$. So an optimal y vector for the lot-sizing instance solves the knapsack instance. So 0–1 KNAPSACK is polynomially reducible to CLS, and as 0–1 KNAPSACK $\in \mathcal{NPC}$, CLS $\in \mathcal{NPC}$. ∎

6.4 CONSEQUENCES OF $\mathcal{P} = \mathcal{NP}$ OR $\mathcal{P} \neq \mathcal{NP}$

Most problems of interest have either been shown to be in $\mathcal{P}$ or in $\mathcal{NPC}$. What is more, nobody has succeeded either in proving that $\mathcal{P} = \mathcal{NP}$ or in showing that $\mathcal{P} \neq \mathcal{NP}$. However, given the huge number of problems in $\mathcal{NPC}$ for which no polynomial algorithm has been found, it is a practical working hypothesis that $\mathcal{P} \neq \mathcal{NP}$.

So how should we interpret the above results and observations?

A first important remark concerns the class $\mathcal{NP}$. Typically, problems in this class have a very large (exponentially large) set of feasible solutions, and these problems can in theory be solved by enumerating the feasible solutions. As we saw in Table 1.1, this is impractical for instances of any reasonable size.

A *pessimist* might say that as most problems appear to be hard (i.e., their decision version lies in $\mathcal{NPC}$), we have no hope of solving instances of large size (because in the worst case we cannot hope to do better than enumeration), and so we should give up.

A *mathematician (optimist)* might set out to become famous by proving that $\mathcal{P} = \mathcal{NP}$

A *mathematician (pessimist)* might set out to become famous by proving that $\mathcal{P} \neq \mathcal{NP}$

A *mathematician (thoughtful)* might decide to ask a different question: Can I find an algorithm that is guaranteed to find a solution "close to optimal" in polynomial time in all cases?

A *probabilist (thoughtful)* might also ask a different question: Can I find an algorithm that runs in polynomial time with high probability and that is guaranteed to find an optimal or "close to optimal" solution with high probability?

An *engineer* would start looking for a heuristic algorithm that produces practically usable solutions.

Your *boss* might say: I don't care a damn about integer programming theory. You just worry about our scheduling problem. Give me a feasible production schedule for tomorrow in which William Brown and Daughters' order is out of the door by 4 P.M.

A *struggling professor* might say: Great. Previously I was trying to develop one algorithm to solve all integer programs, and publishing one paper every two years explaining why I was not succeeding. Now I know that I might as well study each $\mathcal{NP}$ problem individually. As there are thousands of them, I should be able to write twenty papers a year.

Needless to say they are nearly all right. There is no easy and rapid solution, but the problems will not go away, and more and more fascinating and important practical problems are being formulated as integer programs. So in spite of the $\mathcal{NP}$-completness theory, using an appropriate combination of theory, algorithms, experience, and intensive calculation, verifiably good solutions for large instances can and must be found.

Definition 6.7 An optimization problem for which the decision problem lies in $\mathcal{NPC}$ is called $\mathcal{NP}$-*hard*.

The following chapters are devoted to ways to tackle such $\mathcal{NP}$-*hard* problems. First, however, we return briefly to the *Separation Problem* introduced in Chapter 2.

6.5 OPTIMIZATION AND SEPARATION

Here we consider the question of whether there are ties between problems in $\mathcal{P}$. How can we show that a problem is in $\mathcal{P}$? The most obvious way is by finding a polynomial algorithm. We have also seen that another indirect way is by reduction.

There is, however, one general and important result tying together pairs of problems. Put imprecisely, it says:

Given a family of polyhedra associated with a class of problems (such as the convex hulls of the incidence vectors of feasible points $S \subseteq B^n$), the family of optimization problems:

$$\max\{cx : x \in \text{conv}(S)\}$$

is polynomially solvable if and only if the family of separation problems:

Is $x \in \text{conv}(S)$? If not, find an inequality satisfied by all points of S, but cutting off x.

is polynomially solvable.

In other words, the Efficient Optimization and Efficient Separation Properties introduced in Chapter 3 are really equivalent. The other two properties are not exactly equivalent. As we indicated earlier, if a problem has the Efficient Separation Property, it suggests that it may have the Explicit Convex Hull Property. Also if a problem has the Explicit Convex Hull Property, then its linear programming dual may lead to the Strong Dual Property.

6.6 NOTES

6.2 An important step in the development of the distinction between easy and difficult problems is the concept of a certificate of optimality [Edm65a], [Edm65b].

6.3 Cook [Coo71] formally introduced the class $\mathcal{NP}$ and showed the existence of an $\mathcal{NP}$-complete problem. The reduction of many decision versions of integer and combinatorial optimization problems to a $\mathcal{NP}$-complete problem was shown in [Karp72], [Karp75]. The book [GarJoh79] lists an enormous number of $\mathcal{NP}$-complete problems and their reductions. A recent update is [CreKan95].

6.5 The equivalence of optimization and separation is shown in [GroLovSch81], [GroLovSch84]. A more thorough exploration of the equivalence appears in [GroLovSch88]. Other results of importance for integer programming concern the difficulty of finding a short description of all facets for $\mathcal{NP}$-hard problems [PapYan84], and the polynomiality of integer programming with a fixed number of variables [Len83]. Some separation problems are examined in Chapter 9.

For a general book on computational complexity, see [Pap94].

6.7 EXERCISES

1. The 2-*PARTITION* problem is specified by n positive integers $(a_1, \ldots, a_n)$. The problem is to find a subset $S \subset N = \{1, \ldots, n\}$ such that $\sum_{j \in S} a_j =$

$\sum_{j \in N \setminus S} a_j$, or prove that it is impossible. Show that 2-*PARTITION* is polynomially reducible to 0–1 *KNAPSACK*. Does this imply that 2-*PARTITION* is $\mathcal{NP}$-complete?

2. Show that *SATISFIABILITY* is polynomially reducible to *STABLE SET* (Node Packing), and thus that *STABLE SET* is $\mathcal{NP}$-complete, where *STABLE SET* is the problem of finding a maximum weight set of nonadjacent nodes in a graph.

3. Show that *STABLE SET* is polynomially reducible to *SET PACKING*, where *SET PACKING* is the problem of finding a maximum weight set of disjoint columns in a 0–1 matrix.

4. Show that *SET COVERING* is polynomially reducible to *UFL*.

5. Show that *SET COVERING* is polynomially reducible to *DIRECTED STEINER TREE*.

6. Given $D = (V, A)$, c_e for $e \in A$, a subset $F \subseteq A$, and a node $r \in V$, *ARC ROUTING* is the problem of finding a minimum length directed subtour that contains the arcs in F and starts and ends at node r. Show that *TSP* is polynomially reducible to *ARC ROUTING*.

7.* Show that the decision problem associated to *IP* is an integer programming feasibility problem, and is in $\mathcal{NP}$.

8. Consider a 0–1 knapsack set $X = \{x \in B^n : \sum_{j \in N} a_j x_j \le b\}$ with $0 \le a_j \le b$ for $j \in N$ and let $\{x^t\}_{t=1}^T$ be the points of X. With it, associate the bounded polyhedron $\Pi^1 = \{\pi \in R_+^n : x^t \pi \le 1 \text{ for } t = 1, \ldots, T\}$ with extreme points $\{\pi^s\}_{s=1}^S$. Consider a point x^* with $0 \le x_j^* \le 1$ for $j \in N$.

(i) Show that $x^* \in \text{conv}(X)$ if and only if $\min\{\sum_{t=1}^T \lambda_t : x^* \le \sum_{t=1}^T x^t \lambda_t, \lambda \in R_+^T\} = \max\{x^* \pi : \pi \in \Pi^1\} \le 1$.
(ii) Deduce that if $x^* \notin \text{conv}(X)$, then for some $s = 1, \ldots, S$, $\pi^s x^* > 1$.

7

Branch and Bound

7.1 DIVIDE AND CONQUER

Consider the problem:
$$z = \max\{cx : x \in S\}.$$

How can we break the problem into a series of smaller problems that are easier, solve the smaller problems, and then put the information together again to solve the original problem?

Proposition 7.1 *Let $S = S_1 \cup \ldots \cup S_K$ be a decomposition of S into smaller sets, and let $z^k = \max\{cx : x \in S_k\}$ for $k = 1, \ldots, K$. Then $z = \max_k z^k$.*

A typical way to represent such a divide and conquer approach is via an enumeration tree. For instance, if $S \subseteq \{0,1\}^3$, we might construct the enumeration tree shown in Figure 7.1.

Here we first divide S into $S_0 = \{x \in S : x_1 = 0\}$ and $S_1 = \{x \in S : x_1 = 1\}$, then $S_{00} = \{x \in S_0 : x_2 = 0\} = \{x \in S : x_1 = x_2 = 0\}$, $S_{01} = \{x \in S_0 : x_2 = 1\}$, and so on. Note that a leaf of the tree $S_{i_1 i_2 i_3}$ is nonempty if and only if $x = (i_1, i_2, i_3)$ is in S. Thus the leaves of the tree correspond precisely to the points of B^3 that one would examine if one carried out complete enumeration. Note that by convention the tree is drawn upside down with its root at the top.

Another example is the enumeration of all the tours of the traveling salesman problem. First we divide S the set of all tours on 4 cities into $S_{(12)}, S_{(13)}, S_{(14)}$ where $S_{(ij)}$ is the set of all tours containing arc (ij). Then $S_{(12)}$ is divided again into $S_{(12)(23)}$ and $S_{(12)(24)}$, and so on. Note that at the first level we have arbitrarily chosen to branch on the arcs leaving node 1, and at the

91

BRANCH AND BOUND

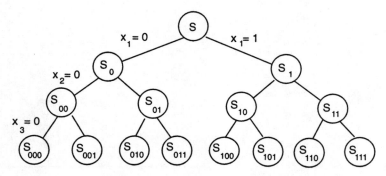

Fig. 7.1 Binary enumeration tree

second level on the arcs leaving node 2 that do not immediately create a subtour with the previous branching arc. The resulting tree is shown in Figure 7.2. Here the six leaves of the tree correspond to the $(n-1)!$ tours shown, where $i_1 i_2 i_3 i_4$ means that the cities are visited in the order i_1, i_2, i_3, i_4, i_1 respectively. Note that this is an example of multiway as opposed to binary branching, where a set can be divided into more than two parts.

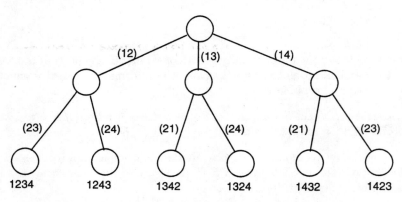

Fig. 7.2 TSP enumeration tree

7.2 IMPLICIT ENUMERATION

We saw in Chapter 1 that complete enumeration is totally impossible for most problems as soon as the number of variables in an integer program, or nodes in a graph exceeds 20 or 30. So we need to do more than just divide indefinitely. How can we use some bounds on the values of $\{z^k\}$ intelligently? First, how can we put together bound information?

$\bar{z} = \max_{k \in \{1,\ldots,K\}} \bar{z}^k$

IMPLICIT ENUMERATION   93

Proposition 7.2 *Let $S = S_1 \cup \ldots \cup S_K$ be a decomposition of S into smaller sets, and let $z^k = \max\{cx : x \in S_k\}$ for $k = 1, \ldots, K$, $\bar{z}^k$ be an upper bound on z^k and $\underline{z}^k$ be a lower bound on z^k. Then $\bar{z} = \max_k \bar{z}^k$ is an upper bound on z and $\underline{z} = \max_k \underline{z}^k$ is a lower bound on z.*

Now we examine three hypothetical examples to see how bound information, or partial information about a subproblem can be put to use. What can be deduced about lower and upper bounds on the optimal value z and which sets need further examination in order to find the optimal value?

Example 7.1 In Figure 7.3 we show a decomposition of S into two sets S_1 and S_2 as well as upper and lower bounds on the corresponding problems.

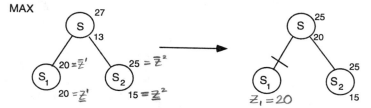

Fig. 7.3 Pruned by optimality

We note first that $\bar{z} = \max_k \bar{z}^k = \max\{20, 25\} = 25$ and $\underline{z} = \max_k \underline{z}^k = \max\{20, 15\} = 20$. So we obtain new upper & lower bounds on master problem.

Second, we observe that as the lower and upper bounds on z_1 are equal, $z_1 = 20$, and there is no further reason to examine the set S_1. Therefore the branch S_1 of the enumeration tree can be pruned *by optimality*. ∎

⟹ $z_1 = 20$, and this subproblem is solved so the corresponding branch is pruned.

Example 7.2 In Figure 7.4 we again decompose S into two sets S_1 and S_2 and show upper and lower bounds on the corresponding problems.

Fig. 7.4 Pruned by bound

We note first that $\bar{z} = \max_k \bar{z}^k = \max\{20, 26\} = 26$ and $\underline{z} = \max_k \underline{z}^k = \max\{18, 21\} = 21$.

Second, we observe that as the optimal value has value at least 21, and the upper bound $\bar{z}^1 = 20$, no optimal solution can lie in the set S_1. Therefore the branch S_1 of the enumeration tree can be pruned *by bound*. ∎

Example 7.3 In Figure 7.5 we again decompose S into two sets S_1 and S_2 with different upper and lower bounds.

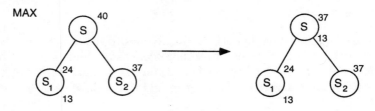

Fig. 7.5 No pruning possible

We note first that $\bar{z} = \max_k \bar{z}^k = \max\{24, 37\} = 37$ and $\underline{z} = \max_k \underline{z}^k = \max\{13, -\} = 13$. Here no other conclusion can be drawn and we need to explore both sets S_1 and S_2 further, ie choose one and branch again.

Based on these examples, we can list at least three reasons that allow us to prune the tree and thus enumerate a large number of solutions implicitly.

(i) Pruning by optimality: $z_t = \{\max cx : x \in S_t\}$ has been solved. (solved the subproblem)

(ii) Pruning by bound: $\bar{z}_t \leq \underline{z}$. (eliminated the subset as a candidate)

(iii) Pruning by infeasiblity: $S_t = \phi$. (the subset is empty so it can't possibly contain the optimum)

If we now ask how the bounds are to be obtained, the reply is no different from in Chapter 2. The primal (lower) bounds are provided by feasible solutions and the dual (upper) bounds by relaxation or duality.

Building an implicit enumeration algorithm based on the above ideas is now in principle a fairly straightforward task. There are, however, many questions that must be addressed before such an algorithm is well-defined. Some of the most important questions are:

What relaxation or dual problem should be used to provide upper bounds? How should one choose between a fairly weak bound that can be calculated very rapidly and a stronger bound whose calculation takes a considerable time?

How should the feasible region be separated into smaller regions $S = S_1 \cup \ldots \cup S_K$? Should one separate into two or more parts? Should one use a fixed a priori rule for dividing up the set, or should the divisions evolve as a function of the bounds and solutions obtained en route?

In what order should the subproblems be examined? Typically there is a list of active problems that have not yet been pruned. Should the next one be chosen on a the basis of last-in first-out, of best/largest upper bound first, or of some totally different criterion?

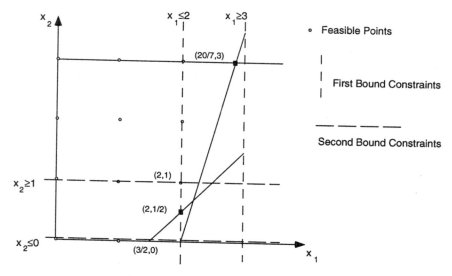

Fig. 7.9   Division of the feasible region

bounds needed to restore the subproblem are stored. Usually one also keeps an optimal or near-optimal basis, so that the linear programming relaxation can be reoptimized rapidly.

Returning to the questions raised earlier, there is no single answer that is best for all instances. One needs to use rules based on a combination of theory, common sense, and practical experimentation. In our example, the question of **how to bound** was solved by using an LP relaxation; **how to branch** was solved by choosing an integer variable that is fractional in the LP solution. However, as there is typically a choice of a set C of several candidates, we need a rule to choose between them. One common choice is *the most fractional variable*:

$$\arg\max_{j \in C} \min[f_j, 1 - f_j]$$

where $f_j = x_j^* - \lfloor x_j^* \rfloor$, so that a variable with fractional value $f_j = \frac{1}{2}$ is best. Other rules are based on the idea of *estimating* the cost of forcing the variable x_j to become integer.

How to choose a node was avoided by making an arbitrary choice. In practice there are several contradictory arguments that can be invoked:

(i) It is only possible to prune the tree significantly with a (primal) feasible solution, giving a hopefully good lower bound. Therefore one should descend as quickly as possible in the enumeration tree to find a first feasible solution. This suggests the use of a *Depth-First Search* strategy. Another argument for

BRANCH AND BOUND

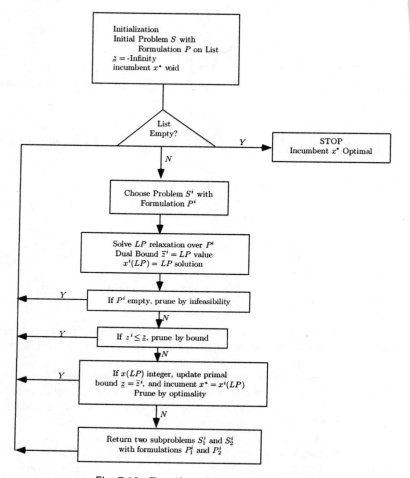

Fig. 7.10 Branch-and-bound flow chart

such a strategy is the observation that it is always easy to resolve the linear programming relaxation when a simple constraint is added, and the optimal basis is available. Therefore passing from a node to one of its immediate descendants is to be encouraged. In the example this would imply that after treating node S_1, the next node treated would be S_{11} or S_{12} rather than S_2.

(ii) To minimize the total number of nodes evaluated in the tree, the optimal strategy is to always choose the active node with the best (largest upper) bound (i.e., choose node s where $\overline{z}_s = \max_t \overline{z}_t$). With such a rule, one will never divide any node whose upper bound $\overline{z}_t$ is less than the optimal value z. This leads to a _Best-Node First_ strategy. In the example of the previous section, this would imply that after treating node S_1, the next node chosen would be S_2 with bound $\frac{59}{7}$ from its predecessor, rather than S_{11} or S_{12} with bound $\frac{15}{2}$.

In practice a compromise between these ideas is often adopted, involving an initial depth-first strategy until at least one feasible solution has been found, followed by a strategy mixing best node and depth first so as to try to prove optimality and also find better feasible solutions.

7.5 USING A BRANCH-AND-BOUND SYSTEM

Commercial branch-and-bound systems for integer and mixed integer programming are essentially as described in the previous section, and the default strategies have been chosen by tuning over hundreds of different problem instances. The basic philosophy is to solve and resolve the linear programming relaxations as rapidly as possible, and if possible to branch intelligently. Given this philosophy, all recent systems contain, or offer,

1. A powerful (automatic) preprocessor, which simplifies the model by reducing the number of constraints and variables, so that the linear programs are easier
2. The simplex algorithm with a choice of pivoting strategies, and an interior point option for solving the linear programs
3. Limited choice of branching and node selection options
4. Use of priorities

and some offer

5. GUB/SOS branching
6. Strong branching
7. Reduced cost fixing
8. Primal heuristics

In this section we briefly discuss those topics requiring user intervention. Preprocessing, which is very important, but automatic, is presented in the (optional) next section. Reduced cost fixing is treated in Exercise 7.7, and primal heuristics are discussed in Chapter 12.

Priorities. Priorities allow the user to tell the system the relative importance of the integer variables. The user provides a file specifying a value (importance) of each integer variable. When it has to decide on a branching variable, the system will choose the highest priority integer variable whose current linear programming value is fractional. At the same time the user can specify a preferred branching direction telling the system which of the two branches to explore first.

GUB Branching. Many models contain generalized upper bound (GUB) or special ordered sets (SOS) of the form

$$\sum_{j=1}^{k} x_j = 1$$

with $x_j \in \{0,1\}$ for $j = 1, \ldots, k$. If the linear programming solution x^* has some of the variables $x_1^*, \ldots, x_k^*$ fractional, then the standard branching rule is to impose $S_1 = S \cap \{x : x_j = 0\}$ and $S_2 = S \cap \{x : x_j = 1\}$ for some $j \in \{1, \ldots, k\}$. However, because of the GUB constraint, $\{x : x_j = 0\}$ leaves $k - 1$ possibilities $\{x : x_i = 1\}_{i \neq j}$ whereas $\{x : x_j = 1\}$ leaves only one possibility. So S_1 is typically a much larger set than S_2, and the tree is *unbalanced*.

GUB branching is designed to provide a more balanced division of S into S_1 and S_2. Specifically the user specifies an ordering of the variables in the GUB set $j_1, \ldots, j_k$, and the branching scheme is then to set

$S_1 = S \cap \{x : x_{j_i} = 0 \ i = 1, \ldots r\}$ and
$S_2 = S \cap \{x : x_{j_i} = 0 \ i = r+1, \ldots k\}$,

where $r = \min\{t : \sum_{i=1}^{t} x_{j_i}^* \geq \frac{1}{2}\}$. In many cases such a branching scheme is much more effective than the standard scheme, and the number of nodes in the tree is significantly reduced.

User Options (a) **Cutoffs.** If the user knows or can construct a good feasible solution to his or her problem, it is very important that its value is passed to the system as the incumbent value to serve as a *cutoff* in the branch and bound.

(b) **Simplex Strategies.** Though the linear programming algorithms are finely tuned, the default strategy will not be best for all classes of problems. Different *simplex pricing* strategies may make a huge difference in running times for a given class of models, so if similar models are resolved repeatedly or the linear programs seem very slow, some experimentation by the user with pricing strategies is permitted. In addition, on very large models, *interior point methods* may be best for the solution of the first linear program. Unfortunately, up to now such methods are still not good for reoptimizing quickly at each node of the tree.

(c) **Strong Branching.** The idea behind strong branching is that on difficult problems it should be worthwhile to do more work to try to choose a better branching variable. The system chooses a set C of basic integer variables that are fractional in the LP solution, branches up and down on each of them in turn, and reoptimizes on each branch either to optimality, or for a specified number of dual simplex pivots. Now for each variable $j \in C$, it has upper bounds z_j^D for the down branch and z_j^U for the up branch. The variable having the largest effect (decrease of the dual bound)

$$j^* = \arg\min_{j \in C} \max[z_j^D, z_j^U]$$

is then chosen, and branching really takes place on this variable. Obviously, solving two *LP*s for each variable in C is costly, so such branching should only be used when the other criteria have been found to be ineffective.

7.5.1 If All Else Fails

What can one do if a particular problem instance turns out to be difficult, meaning that after a certain time

(i) no feasible solution has been found, or

(ii) the gap between the value of the best feasible solution and the value of the dual upper bound is unsatisfactorily large, or

(iii) the system runs out of space because there are too many active nodes in the node list?

Finding Feasible Solutions. This is in general $\mathcal{NP}$-hard. Some systems have simple primal heuristics embedded in them. Also as discussed earlier, using priorities and directions for branching can help. How to find feasible solutions, starting from the LP solution or using explicit problem structure, is the topic of Chapter 12.

Finding Better Dual Bounds. Branch-and-bound algorithms fail very often because the bounds obtained from the linear programming relaxations are too weak. This means that tightening up the formulation of the problem is of crucial importance. Systematic ways to do this are the subject of Chapters 8–11. Specifically the addition of constraints or cuts to improve the formulation is treated in Chapers 8 and 9, leading to the development of a potentially more powerful branch-and-cut algorithm. The Lagrangian relaxation and column generation approaches of Chapters 10 and 11 provide alternative ways to strengthen the formulation by convexifying part of the feasible region.

7.6 PREPROCESSING*

Before solving a linear or integer program, it is natural to check that the formulation is "sensible", and as strong as possible given the information available.

All the commercial branch-and-bound systems carry out such a check, called *preprocessing*. The basic idea is to try to quickly detect and eliminate redundant constraints and variables, and tighten bounds where possible. Then if the resulting linear/integer program is smaller/tighter, it will typically be solved much more quickly. This is especially important in the case of branch-and-bound because tens or hundreds of thousands of linear programs may need to be solved.

First we demonstrate linear programming preprocessing by an example.

Example 7.4 Consider the linear program

$$\begin{aligned}
\max \quad & 2x_1 + x_2 - x_3 \\
& 5x_1 - 2x_2 + 8x_3 \leq 15 \\
& 8x_1 + 3x_2 - x_3 \geq 9 \\
& x_1 + x_2 + x_3 \leq 6 \\
& 0 \leq x_1 \leq 3 \\
& 0 \leq x_2 \leq 1 \\
& 1 \leq x_3.
\end{aligned}$$

Tightening Bounds. Isolating variable x_1 in the first constraint we obtain

$$5x_1 \leq 15 + 2x_2 - 8x_3 \leq 15 + 2 \times 1 - 8 \times 1 = 9$$

where we use the bound inequalities $x_2 \leq 1$ and $-x_3 \leq -1$. Thus we obtain the tightened bound $x_1 \leq \frac{9}{5}$.

Similarly isolating variable x_3, we obtain

$$8x_3 \leq 15 + 2x_2 - 5x_1 \leq 15 + 2 \times 1 - 5 \times 0 = 17,$$

and the tightened bound $x_3 \leq \frac{17}{8}$.

Isolating variable x_2, we obtain

$$2x_2 \geq 5x_1 + 8x_3 - 15 \geq 5 \times 0 + 8 \times 1 - 15 = -7.$$

Here the existing bound $x_2 \geq 0$ is not changed.

Turning to the second constraint, isolating x_1 and using the same approach, we obtain $8x_1 \geq 9 - 3x_2 + x_3 \geq 9 - 3 + 1 = 7$, and an improved lower bound $x_1 \geq \frac{7}{8}$.

No more bounds are changed based on the second or third constraints. However, as certain bounds have been tightened, it is worth passing through the constraints again.

Constraint 1 for x_3 now gives $8x_3 \leq 15 + 2x_2 - 5x_1 \leq 15 + 2 - 5 \times \frac{7}{8} = \frac{101}{8}$. Thus we have the new bound $x_3 \leq \frac{101}{64}$.

Redundant Constraints. Using the latest upper bounds in constraint 3, we see that

$$x_1 + x_2 + x_3 \leq \frac{9}{5} + 1 + \frac{101}{64} < 6,$$

and so this constraint is redundant and can be discarded. The problem is now reduced to

$$
\begin{array}{rllll}
\max & 2x_1 & +x_2 & -x_3 & \\
 & 5x_1 & -2x_2 & +8x_3 & \leq 15 \\
 & 8x_1 & +3x_2 & -x_3 & \geq 9 \\
 & \tfrac{7}{8} \leq x_1 & \leq \tfrac{9}{5}, & 0 \leq x_2 \leq 1, & 1 \leq x_3 \leq \tfrac{101}{64}.
\end{array}
$$

Variable Fixing (by Duality). Considering variable x_2, observe that increasing its value makes all the constraints (other than its bound constraints) less tight. As the variable has a positive objective coefficent, it is advantageous to make the variable as large as possible, and thus set it to its upper bound of 1. (Another way to arrive at a similar conclusion is to write out the LP dual. For the dual constraint corresponding to the primal variable x_2 to be feasible, the dual variable associated with the constraint $x_2 \leq 1$ must be positive. This implies by complemementary slackness that $x_2 = 1$ in any optimal solution.)

Similarly, decreasing x_3 makes the constraints less tight. As the variable has a negative objective coefficient, it is best to make it as small as possible, and thus set it to its lower bound $x_3 = 1$. Finally the *LP* is reduced to the trivial problem

$$\max\{2x_1 : \tfrac{7}{8} \leq x_1 \leq \tfrac{9}{5}\}.$$ ∎

Formalizing the above ideas is straightforward.

Proposition 7.3 *Consider the set* $S = \{x : a_0 x_0 + \sum_{j=1}^{n} a_j x_j \leq b, l_j \leq x_j \leq u_j \text{ for } j = 0, 1, \ldots, n\}$.

(i) Bounds on Variables. If $a_0 > 0$, *then*

$$x_0 \leq (b - \sum_{j: a_j > 0} a_j l_j - \sum_{j: a_j < 0} a_j u_j)/a_0,$$

and if $a_0 < 0$, *then*

$$x_0 \geq (b - \sum_{j: a_j > 0} a_j l_j - \sum_{j: a_j < 0} a_j u_j)/a_0.$$

(ii) Redundancy. The constraint $a_0 x_0 + \sum_{j=1}^{n} a_j x_j \leq b$ *is redundant if*

$$\sum_{j: a_j > 0} a_j u_j + \sum_{j: a_j < 0} a_j l_j \leq b.$$

(iii) Infeasibility. $S = \emptyset$ *if*

$$\sum_{j: a_j > 0} a_j l_j + \sum_{j: a_j < 0} a_j u_j > b.$$

(iv) *Variable Fixing.* For a maximization problem in the form: $\max\{cx : Ax \leq b, l \leq x \leq u\}$, if $a_{ij} \geq 0$ for all $i = 1, \ldots, m$ and $c_j < 0$, then $x_j = l_j$. Conversely if $a_{ij} \leq 0$ for all $i = 1, \ldots, m$ and $c_j > 0$, then $x_j = u_j$.

Turning now to integer programming problems, preprocessing can sometimes be taken a step further. Obviously, if $x_j \in Z^1$ and the bounds l_j or u_j are not integer, we can tighten to

$$\lceil l_j \rceil \leq x_j \leq \lfloor u_j \rfloor.$$

For mixed integer programs with variable upper and lower bound constraints $l_j y_j \leq x_j \leq u_j y_j$ with $y_j \in \{0, 1\}$, it is also important to use the tightest bound information.

For *BIPs* it is common to look for simple "logical" or "boolean" constraints involving only one or two variables, and then either add them to the problem or use them to fix some variables. Again we demonstrate by example.

Example 7.5 Consider the set of constraints involving four 0–1 variables:

$$
\begin{array}{rrrrcr}
7x_1 & +3x_2 & -4x_3 & -2x_4 & \leq & 1 \\
-2x_1 & +7x_2 & +3x_3 & +x_4 & \leq & 6 \\
 & -2x_2 & -3x_3 & -6x_4 & \leq & -5 \\
3x_1 & & -2x_3 & & \geq & -1 \\
& & x & \in & B^4. &
\end{array}
$$

Generating Logical Inequalities. Examining row 1, we see that if $x_1 = 1$, then necessarily $x_3 = 1$, and similarly $x_1 = 1$ implies $x_4 = 1$. This can be formulated with the linear inequalities $x_1 \leq x_3$ and $x_1 \leq x_4$. We see also that the constraint is infeasible if both $x_1 = x_2 = 1$ leading to the constraint $x_1 + x_2 \leq 1$.

Row 2 gives the inequalities $x_2 \leq x_1$ and $x_2 + x_3 \leq 1$.
Row 3 gives $x_2 + x_4 \geq 1$ and $x_3 + x_4 \geq 1$.
Row 4 gives $x_1 \geq x_3$.

Combining Pairs of Logical Inequalities. We consider pairs involving the same variables.

From rows 1 and 4, we have $x_1 \leq x_3$ and $x_1 \geq x_3$, which together give $x_1 = x_3$.

From rows 1 and 2, we have $x_1 + x_2 \leq 1$ and $x_2 \leq x_1$ which together give $x_2 = 0$. Now from $x_2 + x_4 \geq 1$ and $x_2 = 0$, we obtain $x_4 = 1$.

Simplifying. Making the substitutions $x_2 = 0, x_3 = x_1, x_4 = 1$, all four constraints of the feasible region are redundant, and we are left with $x_1 \in \{0, 1\}$, so the only feasible solutions are $(1, 0, 1, 1)$ and $(0, 0, 0, 1)$. ∎

In Exercise 7.10, the reader is asked to formalize the approach taken in this example. The logical inequalities can also be viewed as providing a foretaste of the valid inequalities to be developed in the next chapter.

7.7 NOTES

7.2 The first paper presenting a branch-and-bound algorithm for integer programming is [LanDoi60]. [Litetal63] presents a computationally successful application to the *TSP* problem using an assignment relaxation. [Balas65] developed an algorithm for 0–1 problems using simple tests to obtain dual bounds and check primal feasibility.

7.4 Almost all commercial codes since the 1960s have been linear programming based branch-and-bound codes. The two–way branching scheme commonly used is from [Dak65].

7.5 A discussion of important elements of commercial codes can be found in [Beal79]. GUB/SOS branching is from [BealTom70], probing from [GuiSpi81], and strong branching from [Appetal95]. One important new idea is constraint branching, used for *TSP* problems in [CloNad93], and by [CooRutetal93] in their implementation of basis reduction for integer programming based upon the fundamental paper [Len83]. Recent experiments with various branch-and-bound strategies are reported in [LinSav97].

As solving linear programs forms such an important part of an integer programming algorithm, improvements in solving linear programs are crucial. All recent commercial codes include an interior point algorithm, as for many large linear programs, the latter algorithm is faster than the simplex method. However, because reoptimization with the simplex method is easier than with interior point codes, the simplex method is still used in branch-and-bound. Improving reoptimization with interior point codes is a major challenge for the next few years. See [RooTerVia97] and [Wri97] for recent texts on interior point algorithms. Work on solving integer programs with interior point algorithms is a wide open area [MitTod92],[Mit96].

Knapsack problems, in which the linear programming relaxations can be solved by inspection, have always been treated by specialized codes; see the book [MarTot90].

7.6 Preprocessing is crucially important for the rapid solution of linear programs. Its importance for integer programs is recognized in [BreMitWil73], and discussed more recently in [HofPad91],[Sav94].

7.8 EXERCISES

1. Consider the enumeration tree (minimization problem) in Figure 7.11:

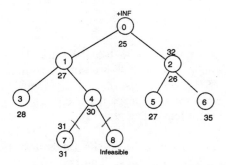

Fig. 7.11 Enumeration tree (min)

(i) Give tightest possible lower and upper bounds on the optimal value z.
(ii) Which nodes can be pruned and which must be explored further?

2. Consider the two-variable integer program:

$$\begin{aligned}
\max \quad & 9x_1 + 5x_2 \\
& 4x_1 + 9x_2 \leq 35 \\
& x_1 \leq 6 \\
& x_1 - 3x_2 \geq 1 \\
& 3x_1 + 2x_2 \leq 19 \\
& x \in \mathbb{Z}_+^2.
\end{aligned}$$

Solve by branch-and-bound graphically and algebraically.

3. Consider the 0–1 knapsack problem:

$$\max\{\sum_{j=1}^n c_j x_j : \sum_{j=1}^n a_j x_j \leq b, x \in B^n\}$$

with $a_j, c_j > 0$ for $j = 1, \ldots, n$.

(i) Show that if $\frac{c_1}{a_1} \geq \ldots \geq \frac{c_n}{a_n} > 0$, $\sum_{j=1}^{r-1} a_j \leq b$ and $\sum_{j=1}^r a_j > b$, the solution of the LP relaxation is $x_j = 1$ for $j = 1, \ldots, r-1$, $x_r = (b - \sum_{j=1}^{r-1} a_j)/a_r$, and $x_j = 0$ for $j > r$.
(ii) Solve the instance

$$\max 17x_1 + 10x_2 + 25x_3 + 17x_4$$
$$5x_1 + 3x_2 + 8x_3 + 7x_4 \leq 12$$
$$x \in B^4$$

by branch-and-bound.

4. Solve the integer knapsack problem:

$$\max 10x_1 + 12x_2 + 7x_3 + \tfrac{3}{2}x_4$$
$$4x_1 + 5x_2 + 3x_3 + 1x_4 \leq 10$$
$$x_1, x_2 \in Z_+^1, x_3, x_4 \in \{0,1\}$$

by branch-and-bound.

5. (i) Solve the $STSP$ instance with $n = 5$ and distance matrix

$$(c_e) = \begin{pmatrix} - & 10 & 2 & 4 & 6 \\ - & - & 9 & 3 & 1 \\ - & - & - & 5 & 6 \\ - & - & - & - & 2 \end{pmatrix}$$

by branch-and-bound using a 1-tree relaxation (see Definition 2.3) to obtain bounds.

(ii) Solve the TSP instance with $n = 4$ and distance matrix

$$(c_{ij}) = \begin{pmatrix} - & 7 & 6 & 3 \\ 3 & - & 6 & 9 \\ 2 & 3 & - & 1 \\ 7 & 9 & 4 & - \end{pmatrix}$$

by branch-and-bound using an assignment relaxation to obtain bounds.
(iii) Describe clearly the branching rules you use in (i) and (ii), and motivate your choice.

6. Using a branch-and-bound system, solve your favorite integer program with different choices of branching and node selection rules, and report on the differences in the running time and the number of nodes in the branch-and-bound tree.

7. *Reduced cost fixing.* Suppose that the linear programming relaxation of an integer program has been solved to optimality, and the objective function is then represented in the form

$$z = \max cx, \, cx = \bar{a}_{00} + \sum_{j \in NB_1} \bar{a}_{0j} x_j + \sum_{j \in NB_2} \bar{a}_{0j}(x_j - u_j)$$

where NB_1 are the nonbasic variables at zero, and NB_2 are the nonbasic variables at their upper bounds u_j, $\bar{a}_{0j} \leq 0$ for $j \in NB_1$, and $\bar{a}_{0j} \geq 0$ for $j \in NB_2$. In addition suppose that a primal feasible solution of value $\underline{z}$ is known. Prove the following: In any optimal solution,

$$x_j \leq \lfloor \tfrac{\bar{a}_{00}-z}{-\bar{a}_{0j}} \rfloor \text{ for } j \in N_1, \text{ and}$$
$$x_j \geq u_j - \lceil \tfrac{\bar{a}_{00}-z}{\bar{a}_{0j}} \rceil \text{ for } j \in N_2.$$

8. Consider a fixed charge network problem:

$$\min\{cx + fy : Nx = b, x \leq uy, x \in R_+^n, y \in Z_+^n\}$$

where N is the node-arc incidence matrix of the network, and b the demand vector. In using priorities, suggest a preferred direction for the variables.

9. Consider the 0–1 problem:
$$\begin{array}{rrrrrrrrrr}
\max & 5x_1 & - & 7x_2 & - & 10x_3 & + & 3x_4 & - & 5x_5 \\
 & x_1 & + & 3x_2 & - & 5x_3 & + & x_4 & + & 4x_5 & \leq & 0 \\
 & -2x_1 & - & 6x_2 & + & 3x_3 & - & 2x_4 & - & 2x_5 & \leq & -4 \\
 & & & 2x_2 & - & 2x_3 & - & x_4 & + & x_5 & \leq & -2 \\
 & & & & & & & x & \in & B^5.
\end{array}$$

Simplify using logical inequalties.

10. *Logical.* Given a set in 0–1 variables

$$X = \{x \in B^n : \sum_{j=1}^{n} a_j x_j \leq b\}$$

with $a_j \geq 0$ for $j = 1, \ldots, n$, under what conditions is

(i) the set X empty?
(ii) the constraint $\sum_{j=1}^{n} a_j x_j \leq b$ redundant?
(iii) the constraint $x_j = 0$ valid?
(iv) the constraint $x_i + x_j \leq 1$ valid?

Apply these rules to the first constraint in Exercise 9.

11. Prove Proposition 7.3 concerning preprocessing.

12. Let

$$X = \{x \in B^n : \sum_{j=1}^{n} a_j x_j \leq b\}$$

with $a_1 \geq a_2 \geq \ldots \geq a_n \geq 0$ and $b \geq 0$. The idea is to write each such set in some simple canonical form. For example, for $x \in B^3$, $12x_1 + 8x_2 + 3x_1 \leq 14$ is equivalent to $2x_1 + 1x_2 + 1x_3 \leq 2$.

(i) When $n = 2$, how many distinct knapsack sets are there? Write them out in a canonical form with integral coefficients and $1 = a_1 \geq a_2$.

(ii) Repeat for $n = 3$ with $a_1 \leq 2$.

13*. Some small integer programs are very difficult for mixed integer programming systems. Try to find a feasible solution to the integer equality knapsack: $\{x \in Z_+^n : \sum_{j=1}^n a_j x_j = b\}$ with $a = (12228, 36679, 36682, 48908, 61139, 73365)$ and $b = 89716837$.

14. Suppose that $P^i = \{x \in R^n : A^i x \leq b^i\}$ for $i = 1, \ldots, m$ and that $C_k \subseteq \{1, \ldots, m\}$ for $k = 1, \ldots, K$. A *disjunctive program* is a problem of the form
$$\max\{cx : x \in \cup_{i \in C_k} P^i \text{ for } k = 1, \ldots, K\}.$$
Show how the following can be modeled as disjunctive programs:

(i) a 0–1 integer program.
(ii) a linear complementarity problem: $w = q + Mz, w, z \in R_+^m, w_j z_j = 0$ for $j = 1, \ldots, m$.
(iii) a single machine sequencing problem with job processing times p_j, and variables t_j representing the start time of job j for $j = 1, \ldots, n$.
(iv) the nonlinear expression $z = \max\{3x_1 + 2x_2 - 3, 9x_1 - 4x_2 + 6\}$.
(v) the constraint: if $x \in R_+^1$ is positive, then x lies between 20 and 50, and is a multiple of 5.

15. Devise a branch-and-bound algorithm for a disjunctive program.

8
Cutting Plane Algorithms

8.1 INTRODUCTION

Here we consider the general integer program:

$$(IP) \qquad \max\{cx : x \in X\}$$

where $X = \{x : Ax \leq b, x \in Z_+^n\}$.

Proposition 8.1 $conv(X) = \{x : \tilde{A}x \leq \tilde{b}, x \geq 0\}$ *is a polyhedron.*

This result, already presented in Chapter 1, tells us that we can, in theory, reformulate problem IP as the linear program:

$$(LP) \qquad \max\{cx : \tilde{A}x \leq \tilde{b}, x \geq 0\}$$

and then for any value of c, an optimal extreme point solution of LP is an optimal solution of IP. The same result holds for mixed integer programs with $X = \{(x,y) \in R_+^n \times Z_+^p : Ax + Gy \leq b\}$ provided the data A, G, b are rational.

In Chapter 3 we have seen several problems, including the assignment problem and the spanning tree problem, for which we have given an explicit description of $conv(X)$. However, unfortunately for $\mathcal{NP}$-hard problems, there is almost no hope of finding a "good" description. Given an instance of an $\mathcal{NP}$-hard problem, the goal in this chapter is to find effective ways to try and approximate $conv(X)$ for the given instance.

The fundamental concept that we have already used informally is that of a valid inequality.

114 CUTTING PLANE ALGORITHMS

Definition 8.1 An inequality $\pi x \leq \pi_0$ is a *valid inequality* for $X \subseteq R^n$ if $\pi x \leq \pi_0$ for all $x \in X$.

If $X = \{x \in Z^n : Ax \leq b\}$ and conv$(X) = \{x \in R^n : \tilde{A}x \leq \tilde{b}\}$, the constraints $a^i x \leq b_i$ and $\tilde{a}^i x \leq \tilde{b}_i$ are clearly valid inequalities for X.

The two questions that immediately spring to mind are

(i) Which are the "good" or useful valid inequalities? and

(ii) If we know a set or family of valid inequalities for a problem, how can we use them in trying to solve a particular instance?

8.2 SOME SIMPLE VALID INEQUALITIES

First we present some examples of valid inequalities. The first type, logical inequalities, have already been seen in Example 7.5 in looking at preprocessing.

Example 8.1 A Pure 0–1 Set. Consider the 0–1 knapsack set:

$$X = \{x \in B^5 : 3x_1 - 4x_2 + 2x_3 - 3x_4 + x_5 \leq -2\}.$$

If $x_2 = x_4 = 0$, the lhs (left-hand side) $= 3x_1 + 2x_3 + x_5 \geq 0$ and the rhs (right-hand side) $= -2$, which is impossible. So all feasible solutions satisfy the valid inequality $x_2 + x_4 \geq 1$.

If $x_1 = 1$ and $x_2 = 0$, the lhs $= 3 + 2x_3 - 3x_4 + x_5 \geq 3 - 3 = 0$ and the rhs $= -2$, which is impossible, so $x_1 \leq x_2$ is also a valid inequality. ∎

Example 8.2 A Mixed 0–1 Set. Consider the set:

$$X = \{(x, y) : x \leq 9999y, 0 \leq x \leq 5, y \in B^1\}.$$

It is easily checked that the inequality

$$x \leq 5y$$

is valid because $X = \{(0,0), (x,1) \text{ with } 0 \leq x \leq 5\}$. As X only involves two variables, it is possible to represent X graphically, so it is easy to check that the addition of the inequality $x \leq 5y$ gives us the convex hull of X.

Such constraints arise often. For instance, in the capacitated facility location problem one has the feasible region:

$$\sum_{i \in M} x_{ij} \leq b_j y_j \text{ for } j \in N$$

$$\sum_{j \in N} x_{ij} = a_i \text{ for } i \in M$$

$$x_{ij} \geq 0 \text{ for } i \in M, j \in N, y_j \in \{0,1\} \text{ for } j \in N.$$

SOME SIMPLE VALID INEQUALITIES 115

All feasible solutions satisfy $x_{ij} \leq b_j y_j$ and $x_{ij} \leq a_i$ with $y_j \in B^1$. This is precisely the situation above leading to the family of valid inequalities $x_{ij} \leq \min\{a_i, b_j\} y_j$. ∎

Example 8.3 A Mixed Integer Set. Consider the set

$$X = \{(x,y) : x \leq 10y, 0 \leq x \leq 14, y \in Z_+^1\}. \quad (x \in \mathbb{R})$$

It is not difficult to verify the validity of the inequality $x \leq 6 + 4y$, or written another way, $x \leq 14 - 4(2-y)$. In Figure 8.1 we represent X graphically, and see that the addition of the inequality $x \leq 6 + 4y$ gives the convex hull of X.

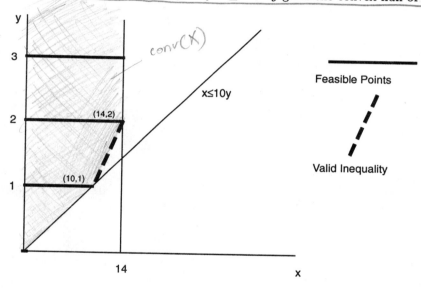

Fig. 8.1 Mixed integer inequality

For the general case, when C does not divide b, and

$$X = \{(x,y) : x \leq Cy, 0 \leq x \leq b, y \in Z_+^1\},$$

one obtains the valid inequality $x \leq b - \gamma(K - y)$ where $K = \lceil \frac{b}{C} \rceil$ and $\gamma = b - (\lceil \frac{b}{C} \rceil - 1)C$. ∎

Example 8.4 A Combinatorial Set: Matchings. Consider the X of incidence vectors of matchings:

$$\sum_{e \in \delta(i)} x_e \leq 1 \text{ for } i \in V \qquad (8.1)$$

$$x \in Z_+^{|E|} \qquad (8.2)$$

where $\delta(i) = \{e \in E : e = (i,j) \text{ for some } j \in V\}$.

Take a set $T \subseteq V$ of nodes of odd cardinality. As the edges of a matching are disjoint, the number of edges of a matching having both endpoints in T is at most $\frac{|T|-1}{2}$. Therefore

$$\sum_{e \in E(T)} x_e \leq \frac{|T|-1}{2} \qquad (8.3)$$

is a valid inequality for X if $|T| \geq 3$ and $|T|$ is odd. ∎

Example 8.5 Integer Rounding. Consider the integer region $X = P \cap Z^4$ where

$$P = \{x \in R_+^4 : 13x_1 + 20x_2 + 11x_3 + 6x_4 \geq 72\}.$$

Dividing by 11 gives the valid inequality for P:

$$\tfrac{13}{11}x_1 + \tfrac{20}{11}x_2 + x_3 + \tfrac{6}{11}x_4 \geq 6\tfrac{6}{11}.$$

As $x \geq 0$, rounding up the coefficients on the left to the nearest integer gives $2x_1 + 2x_2 + x_3 + x_4 \geq \tfrac{13}{11}x_1 + \tfrac{20}{11}x_2 + x_3 + \tfrac{6}{11}x_4 \geq 6\tfrac{6}{11}$, and so we get a weaker valid inequality for P:

$$2x_1 + 2x_2 + x_3 + x_4 \geq 6\tfrac{6}{11}.$$

As x is integer and all the coefficients are integer, the lhs must be integer. An integer that is greater than or equal to $6\tfrac{6}{11}$ must be at least 7, and so we can round the rhs up to the nearest integer giving the valid inequality for X:

$$2x_1 + 2x_2 + x_3 + x_4 \geq 7.$$

∎

Such regions arise in many problems. Consider, for instance, a *Generalized Transportation Problem* where the problem is to satisfy the demand d_j of client j using trucks of different types. A truck of type i has capacity C_i, there are a_i of them available, and the cost if a truck of type i is sent to client j is c_{ij}. The resulting integer program is:

$$\min \sum_{i=1}^{n} \sum_{j=1}^{n} c_{ij} x_{ij}$$

$$\sum_{i=1}^{n} C_i x_{ij} \geq d_j \text{ for } j = 1, \ldots, n$$

$$\sum_{j=1}^{n} x_{ij} \leq a_i \text{ for } i = 1, \ldots, m$$

$$x \in Z_+^{mn},$$

where each demand constraint gives rise to a set of the form X.

Example 8.6 Mixed Integer Rounding. Consider the same example as above with the addition of a continuous variable. Let $X = P \cap (Z^4 \times R^1)$ where

$$P = \{(y, s) \in R_+^4 \times R_+^1 : 13y_1 + 20y_2 + 11y_3 + 6y_4 + s \geq 72\}.$$

In terms of the generalized transportation model, there are four types of truck available to satisfy demand, but it is also possible to satisfy demand from an alternative source. Dividing by 11 gives

$$\tfrac{13}{11}y_1 + \tfrac{20}{11}y_2 + y_3 + \tfrac{6}{11}y_4 \geq \tfrac{72-s}{11},$$

suggesting that there is a valid inequality

$$2y_1 + 2y_2 + y_3 + y_4 + \alpha s \geq 7 \text{ for some } \alpha. \tag{8.4}$$

Looking at the rhs term $\tfrac{72-s}{11}$, we see that the rhs $\lceil \tfrac{72-s}{11} \rceil$ decreases from 7 to 6 at the critical value $s = 6$, indicating the value $\alpha = \tfrac{1}{6}$. Inequality (8.4) turns out to be valid for values of $\alpha \geq \tfrac{1}{6}$, and later we will see that it can even be strengthened a little, giving:

$$\tfrac{3}{2}y_1 + 2y_2 + y_3 + y_4 + \tfrac{1}{6}s \geq 7. \qquad \blacksquare$$

8.3 VALID INEQUALITIES

To understand how to generate valid inequalities for integer programs, it is first necessary to understand valid inequalities for polyhedra (or linear programs).

8.3.1 Valid Inequalities for Linear Programs

So the first question is: When is the inequality $\pi x \leq \pi_0$ valid for $P = \{x : Ax \leq b, x \geq 0\}$?

Proposition 8.2 $\pi x \leq \pi_0$ *is valid for* $P = \{x : Ax \leq b, x \geq 0\} \neq \emptyset$ *if and only if there exist* $u \geq 0, v \geq 0$ *such that* $uA - v = \pi$ *and* $ub \leq \pi_0$, *or alternatively there exists* $u \geq 0$ *such that* $uA \geq \pi$ *and* $ub \leq \pi_0$.

Proof. By linear programming duality, $\max\{\pi x : x \in P\} \leq \pi_0$ if and only if $\min\{ub : uA - v = \pi, u \geq 0, v \geq 0\} \leq \pi_0$. ∎

8.3.2 Valid Inequalities for Integer Programs

Now we consider the feasible region of an integer program:

$$\{x : Ax \leq b, x \in Z_+^n\}$$

and ask the same question.

Surprisingly, the complete answer is in some sense given in the following very simple observation.

Proposition 8.3 *Let* $X = \{y \in Z^1 : y \leq b\}$, *then the inequality* $y \leq \lfloor b \rfloor$ *is valid for* X.

We have already used this idea in Example 8.5. We now give two more examples.

Example 8.4 (cont) Valid Inequalities for Matching. Here we give an alternative algebraic justification for the validity of inequality (8.3) that can be broken up into three steps.

(i) Take a nonnegative linear combination of the constraints (8.1) with weights $u_i = \frac{1}{2}$ for $i \in T$ and $u_i = 0$ for $i \in V \setminus T$. This gives the valid inequality:

$$\sum_{e \in E(T)} x_e + \frac{1}{2} \sum_{e \in \delta(T, V \setminus T)} x_e \leq \frac{|T|}{2}.$$

(ii) Because $x_e \geq 0$, $\sum_{e \in E(T)} x_e \leq \sum_{e \in E(T)} x_e + \frac{1}{2} \sum_{e \in \delta(T, V \setminus T)} x_e$, and so

$$\sum_{e \in E(T)} x_e \leq \frac{|T|}{2}$$

is a valid inequality.

(iii) Because $x \in Z^n$, the lhs $\sum_{e \in E(T)} x_e$ must be an integer, and so one can replace the rhs value by the largest integer less than or equal to the rhs value. So

$$\sum_{e \in E(T)} x_e \leq \lfloor \frac{|T|}{2} \rfloor$$

is a valid inequality. ∎

Example 8.7 Valid Inequalities for an Integer Program. An identical approach can be used to derive valid inequalities for any integer programming region. Let $X = P \cap Z^n$ be the set of integer points in P where P is given by:

$$\begin{aligned} 7x_1 - 2x_2 &\leq 14 \\ x_2 &\leq 3 \\ 2x_1 - 2x_2 &\leq 3 \\ x &\geq 0. \end{aligned}$$

VALID INEQUALITIES

(i) First combining the constraints with nonnegative weights $u = (\frac{2}{7}, \frac{37}{63}, 0)$, we obtain the valid inequality for P

$$2x_1 + \frac{1}{63}x_2 \leq \frac{121}{21}.$$

(ii) Reducing the coefficients on the left-hand side to the nearest integer gives the valid inequality for P:

$$2x_1 + 0x_2 \leq \frac{121}{21}.$$

(iii) Now as the left-hand side is integral for all points of X, we can reduce the rhs to the nearest integer, and we obtain the valid inequality for X:

$$2x_1 \leq \lfloor \frac{121}{21} \rfloor = 5.$$

Observe that if we repeat the procedure, and use a weight of $\frac{1}{2}$ on this last constraint, we obtain the tighter inequality $x_1 \leq \lfloor \frac{5}{2} \rfloor = 2$. ∎

Now it is easy to describe the general procedure that we have been using.

Chvátal-Gomory procedure to construct a valid inequality for the set $X = P \cap Z^n$, where $P = \{x \in R_+^n : Ax \leq b\}$, A is an $m \times n$ matrix with columns $\{a_1, a_2, \ldots, a_n\}$, and $u \in R_+^m$:

$P = \{x \in R_+^n : \sum_{j=1}^{n} a_j x_j \leq b\}$

(i) the inequality

$$\sum_{j=1}^{n} u a_j x_j \leq ub \qquad u^T a_1 x_1 + u^T a_2 x_2 + \cdots + u^T a_n x_n \leq u^T b$$

is valid for P as $u \geq 0$ and $\sum_{j=1}^{n} a_j x_j \leq b$,

(ii) the inequality

$$\sum_{j=1}^{n} \lfloor u a_j \rfloor x_j \leq ub$$

is valid for P as $x \geq 0$,

(iii) the inequality

$$\sum_{j=1}^{n} \lfloor u a_j \rfloor x_j \leq \lfloor ub \rfloor$$

is valid for X as x is integer, and thus $\sum_{j=1}^{n} \lfloor u a_j \rfloor x_j$ is integer.

The surprising fact is that this simple procedure is sufficient to generate all valid inequalities for an integer program.

Theorem 8.4 *Every valid inequality for X can be obtained by applying the Chvátal-Gomory procedure a finite number of times.*

Proof.* We present a proof for the 0-1 case. Thus let $P = \{x \in R^n : Ax \leq b, 0 \leq x \leq 1\} \neq \emptyset$, $X = P \cap Z^n$, and suppose that $\pi x \leq \pi_0$ with π, π_0 integral is a valid inequality for X. We will show that this inequality is a C-G inequality, or in other words that it can be obtained by applying the Chvátal-Gomory procedure a finite number of times.

Claim 1 The inequality $\pi x \leq \pi_0 + t$ is a valid inequality for P for some $t \in Z^1_+$.
Proof. $z_{LP} = \max\{cx : x \in P\}$ is bounded as P is bounded. Take $t = \lceil z_{LP} \rceil - \pi_0$.

Claim 2 There exists a sufficiently large integer M that the inequality

$$\pi x \leq \pi_0 + M \sum_{j \in N^0} x_j + M \sum_{j \in N^1} (1 - x_j) \tag{8.5}$$

is valid for P for every partition (N^0, N^1) of N.
Proof. It suffices to show that the inequality is satisfied at all vertices x^* of P. If $x^* \in Z^n$, then the inequality $\pi x \leq \pi_0$ is satisfied, and so (8.5) is satisfied. Otherwise there exists $\alpha > 0$ such that $\sum_{j \in N^0} x_j^* + \sum_{j \in N^1} (1 - x_j^*) \geq \alpha$ for all partitions (N^0, N^1) of N and all non-integral extreme points x^* of P. Taking $M \geq \frac{t}{\alpha}$, it follows that for all extreme points of P,

$$\pi x^* \leq \pi_0 + t \leq \pi_0 + M \sum_{j \in N^0} x_j^* + M \sum_{j \in N^1} (1 - x_j^*).$$

Claim 3 If $\pi x \leq \pi_0 + \tau + 1$ is a C-G inequality for X with $\tau \in Z^1_+$, then

$$\pi x \leq \pi_0 + \tau + \sum_{j \in N^0} x_j + \sum_{j \in N^1} (1 - x_j) \tag{8.6}$$

is a C-G inequality for X.
Proof. Take the inequality $\pi x \leq \pi_0 + \tau + 1$ with weight $(M-1)/M$ and the inequality (8.5) with weight $1/M$. The resulting C-G inequality is (8.6).

Claim 4 If

$$\pi x \leq \pi_0 + \tau + \sum_{j \in T^0 \cup \{p\}} x_j + \sum_{j \in T^1} (1 - x_j) \tag{8.7}$$

and

$$\pi x \leq \pi_0 + \tau + \sum_{j \in T^0} x_j + \sum_{j \in T^1 \cup \{p\}} (1 - x_j) \tag{8.8}$$

are C-G inequalities for X where (T^0, T^1) is any partition of $\{1,\ldots,p-1\}$, then

$$\pi x \leq \pi_0 + \tau + \sum_{j \in T^0} x_j + \sum_{j \in T^1}(1 - x_j) \tag{8.9}$$

is a C-G inequality for X.
Proof. Take the inequalities (8.7) and (8.8) with weights 1/2 and the resulting C-G inequality is (8.9).

Claim 5 If
$$\pi x \leq \pi_0 + \tau + 1$$
is a C-G inequality for X, then
$$\pi x \leq \pi_0 + \tau$$
is a C-G inequality for X.
Proof. Apply Claim 4 successively for $p = n, n-1, \ldots, 1$ with all partitions (T^0, T^1) of $\{1, \ldots, p-1\}$.

Finally starting with $\tau = t-1$ and using Claim 5 for $\tau = t-1, \cdots, 0$ establishes that $\pi x \leq \pi_0$ is a C-G inequality. ∎

For inequalities generated by other arguments, it is sometimes interesting to see how easily they are generated as C-G inequalities, see Exercise 8.15.

Now that we have seen a variety of both ad hoc and general ways to derive valid inequalities, we turn to the important practical question of how to use them.

8.4 A PRIORI ADDITION OF CONSTRAINTS

In discussing branch-and-bound we saw that preprocessing was a first step in tightening a formulation. Here we go a step further. The idea here is to examine the initial formulation $P = \{x : Ax \leq b, x \geq 0\}$ with $X = P \cap Z^n$, find a set of valid inequalities $Qx \leq q$ for X, add these to the formulation immediately giving a new formulation $P' = \{x : Ax \leq b, Qx \leq q, x \geq 0\}$ with $X = P' \cap Z^n$. Then one can apply one's favorite algorithm, Branch-and-Bound or whatever, to formulation P'.

Advantages. One can use standard branch-and-bound software. If the valid inequalities are well chosen so that formulation P' is significantly smaller than P, the bounds should be improved and hence the branch-and-bound algorithm should be more effective. In addition the chances of finding feasible integer solutions in the course of the algorithm should increase.

Disadvantages. Often the family of valid inequalities one would like to add is enormous. In such cases either the linear programs become very big and take a long time to solve, or it becomes impossible to use standard branch-and-bound software because there are too many constraints.

How can one start looking for valid inequalities a priori? In many instances the feasible region X can be written naturally as the intersection of two or more sets with structure, that is, $X = X^1 \cap X^2$. *Decomposing* or *concentrating on one of the sets at a time* may be a good idea.

For instance, we may know that the optimization problem over the set $X^2 = P^2 \cap Z^n$ is easy. Then we may be able to find an explicit description of $P'^2 = \operatorname{conv}(P^2 \cap Z^n)$. In this case we can replace the initial formulation $P^1 \cap P^2$ by an improved formulation $P' = P^1 \cap P'^2$.

Whether the optimization problem over X^2 is easy or not, one may be able to take advantage of the structure to find valid inequalities for X^2, which allow us to improve its formulation from P^2 to $P'^2 \subset P^2$, and thereby again provide an improved formulation $P' = P^1 \cap P'^2$ for X.

Example 8.8 Uncapacitated Facility Location. Take the "weak" formulation used in the 1950s and 1960s:

$$\sum_{j=1}^n x_{ij} = 1 \text{ for } i = 1, \ldots, m$$

$$\sum_{i=1}^m x_{ij} \leq m y_j \text{ for } j = 1, \ldots, n$$

$$x_{ij} \geq 0 \text{ for } i = 1, \ldots, m, \ j = 1, \ldots, n$$

$$0 \leq y_j \leq 1 \text{ for } j = 1, \ldots, n.$$

Let X_j be the set of points in the polyhedron P_j :

$$\sum_{i=1}^m x_{ij} \leq m y_j$$

$$x_{ij} \geq 0 \text{ for } i = 1, \ldots, m$$

$$0 \leq y_j \leq 1,$$

with y_j integer. The convex hull P'_j of the set X_j is given by

$$x_{ij} \leq y_j \text{ for } i = 1, \ldots, m$$

$$x_{ij} \geq 0 \text{ for } i = 1, \ldots, m$$

$$0 \leq y_j \leq 1.$$

Now the reformulation obtained by replacing P_j by P'_j for $j = 1, \ldots, n$ is the "strong" formulation P':

$$\sum_{j=1}^{n} x_{ij} = 1 \text{ for } i = 1, \ldots, m$$
$$x_{ij} \leq y_j \text{ for } i = 1, \ldots, m \ j = 1, \ldots, n$$
$$x_{ij} \geq 0 \text{ for } i = 1, \ldots, m, \ j = 1, \ldots, n$$
$$0 \leq y_j \leq 1 \text{ for } j = 1, \ldots, n.$$

The strong formulation is now commonly used for this problem because the bound it provides is much stronger than that given by the weak formulation, and the linear programming relaxation has solutions that are close to being integral. ∎

Example 8.9 Constant Capacity Lot-Sizing. Using the same notation as for ULS introduced in Section 1.4, a basic formulation of the feasible region X is

$$s_{t-1} + x_t = d_t + s_t \text{ for } t = 1, \ldots, n$$
$$x_t \leq Cy_t \text{ for } t = 1, \ldots, n$$
$$s_0 = s_n = 0, s \in R_+^{n+1}, x \in R^n, y \in B^n.$$

We derive two families of inequalities for X. First consider any point $(x, s, y) \in X$. First as $s_{t-1} \geq 0$, the inequality $x_t \leq d_t + s_t$ clearly holds. Along with $x_t \leq Cy_t$ and $y_t \in \{0, 1\}$, it is then not difficult to show that $x_t \leq d_t y_t + s_t$ is valid for X. (Note that without the variable s_t, this is precisely the mixed 0-1 inequality of Example 8.2).

Second, summing the flow conservation constraints, and using $s_t \geq 0$, we get the inequality $\sum_{i=1}^{t} x_i \geq \sum_{i=1}^{t} d_i$. Then using $x_i \leq Cy_i$ gives $C \sum_{i=1}^{t} y_i \geq \sum_{i=1}^{t} d_i$, or $\sum_{i=1}^{t} y_i \geq (\sum_{i=1}^{t} d_i)/C$. Now as $\sum_{i=1}^{t} y_i$ is integral, we can use Chvátal-Gomory integer rounding to obtain the valid inequality

$$\sum_{i=1}^{t} y_i \geq \lceil \frac{\sum_{i=1}^{t} d_i}{C} \rceil.$$

Adding just these $2n$ inequalities significantly strengthens the formulation of this problem. ∎

8.5 AUTOMATIC REFORMULATION OR CUTTING PLANE ALGORITHMS

Suppose that $X = P \cap Z^n$ and that we know a family $\mathcal{F}$ of valid inequalities $\pi x \leq \pi_0, (\pi, \pi_0) \in \mathcal{F}$ for X. In many cases $\mathcal{F}$ contains too many inequalities

(2^n or more) for them to be added a priori. Also given a specific objective function, one is not really interested in finding the complete convex hull, but one wants a good approximation to it in the neighborhood of the optimal solution.

We now describe a basic cutting plane algorithm for (IP), $\max\{cx : x \in X\}$, that generates "useful" inequalities from $\mathcal{F}$.

Cutting Plane Algorithm

Initialization. Set $t = 0$ and $P^0 = P$.
Iteration t. Solve the linear program:

$$\overline{z}^t = \max\{cx : x \in P^t\}.$$

Let x^t be an optimal solution.
If $x^t \in Z^n$, stop. x^t is an optimal solution for IP.
If $x^t \notin Z^n$, solve the separation problem for x^t and the family $\mathcal{F}$.
If an inequality $(\pi^t, \pi_0^t) \in \mathcal{F}$ is found with $\pi^t x^t > \pi_0^t$ so that it cuts off x^t, set $P^{t+1} = P^t \cap \{x : \pi^t x \leq \pi_0^t\}$, and augment t. Otherwise stop.

If the algorithm terminates without finding an integral solution for IP,

$$P^t = P \cap \{x : \pi^i x \leq \pi_0^i \ \ i = 1, \ldots, t\}$$

is an improved formulation that can be input to a branch-and-bound algorithm. It should also be noted that in practice it is often better to add several violated cuts at each iteration, and not just one at a time.

In the next section we look at a specific implementation of this algorithm.

8.6 GOMORY'S FRACTIONAL CUTTING PLANE ALGORITHM

Here we consider the integer program:

$$\max\{cx : Ax = b, x \geq 0 \text{ and integer}\}.$$

The idea is to first solve the associated linear programming relaxation and find an optimal basis, choose a basic variable that is not integer, and then generate a Chvátal-Gomory inequality on the constraint associated with this basic variable so as to cut off the linear programming solution. We suppose, given an optimal basis, that the problem is rewritten in the form:

$$\max \overline{a}_{00} + \sum_{j \in NB} \overline{a}_{0j} x_j$$
$$x_{B_u} + \sum_{j \in NB} \overline{a}_{uj} x_j = \overline{a}_{u0} \text{ for } u = 1, \ldots, m$$
$$x \geq 0 \text{ and integer}$$

GOMORY'S FRACTIONAL CUTTING PLANE ALGORITHM

with $\bar{a}_{0j} \leq 0$ for $j \in NB$ and $\bar{a}_{u0} \geq 0$ for $u = 1, \ldots, m$, where NB is the set of nonbasic variables.

If the basic optimal solution x^* is not integer, there exists some row u with $\bar{a}_{u0} \notin Z^1$. Choosing such a row, the Chvátal-Gomory cut for row u is

$$x_{B_u} + \sum_{j \in NB} \lfloor \bar{a}_{uj} \rfloor x_j \leq \lfloor \bar{a}_{u0} \rfloor. \qquad (8.10)$$

note, to do the cut we switch the equality in Tableau to inequality, then round down

Rewriting this inequality by eliminating x_{B_u} gives

$$\sum_{j \in NB} (\bar{a}_{uj} - \lfloor \bar{a}_{uj} \rfloor) x_j \geq \bar{a}_{u0} - \lfloor \bar{a}_{u0} \rfloor$$

$X_{B_u} = \bar{a}_{u0} - \sum_{j \in NB} \bar{a}_{uj} X_j$

or

$$\sum_{j \in NB} f_{uj} x_j \geq f_{u0} > 0 \qquad (8.11)$$

where $f_{uj} = \bar{a}_{uj} - \lfloor \bar{a}_{uj} \rfloor$ for $j \in NB$, and $f_{u0} = \bar{a}_{u0} - \lfloor \bar{a}_{u0} \rfloor$.

By the definitions and the choice of row u, $0 \leq f_{uj} < 1$ and $0 < f_{u0} < 1$. As $x_j^* = 0$ for all nonbasic variables $j \in NB$ in the optimal LP solution, this inequality cuts off x^*. It is also important to observe that the difference between the left- and right-hand sides of the Chvátal-Gomory inequality (8.10), and hence also of (8.11), is integral when x is integral, so that when (8.11) is rewritten as an equation:

$$s = -f_{u0} + \sum_{j \in NB} f_{uj} x_j,$$

the slack variable s is a nonnegative integer variable.

Example 8.10 Consider the integer program

$$z = \max \begin{bmatrix} 4x_1 & - & x_2 \\ 7x_1 & - & 2x_2 & \leq & 14 \\ & & x_2 & \leq & 3 \\ 2x_1 & - & 2x_2 & \leq & 3 \\ x_1, & & x_2 & \geq & 0 \text{ and integer.} \end{bmatrix}$$

Adding slack variables x_3, x_4, x_5, observe that as the constraint data is integer, the slack variables must also take integer values. Now solving as a linear program gives:

$$z = \max \begin{bmatrix} \frac{59}{7} & & -\frac{4}{7}x_3 & -\frac{1}{7}x_4 & & & \\ x_1 & & +\frac{1}{7}x_3 & +\frac{2}{7}x_4 & & = & \frac{20}{7} \\ & x_2 & & +x_4 & & = & 3 \\ & & -\frac{2}{7}x_3 & +\frac{10}{7}x_4 & +x_5 & = & \frac{23}{7} \\ x_1, & x_2, & x_3, & x_4, & x_5 & \geq & 0 \text{ and integer.} \end{bmatrix}$$

The optimal linear programming solution is $x = (\frac{20}{7}, 3, 0, 0, \frac{23}{7}) \notin \mathbb{Z}_+^5$, so we use the first row, in which the basic variable x_1 is fractional, to generate the cut:

$$\tfrac{1}{7}x_3 + \tfrac{2}{7}x_4 \geq \tfrac{6}{7} \qquad \tfrac{20}{7} - \lfloor \tfrac{20}{7} \rfloor = \tfrac{6}{7}$$

or

$$s = -\tfrac{6}{7} + \tfrac{1}{7}x_3 + \tfrac{2}{7}x_4$$

with $s, x_3, x_4 \geq 0$ and integer.

Adding this cut, and reoptimizing leads to the new optimal tableau

$$\begin{array}{rrrrrrrl}
z = \max \big[\tfrac{15}{2} & & & & -\tfrac{1}{2}x_5 & -3s & & \big] \\
& x_1 & & & & +s & & = 2 \\
& & x_2 & & -\tfrac{1}{2}x_5 & +s & & = \tfrac{1}{2} \\
& & & x_3 & -x_5 & -5s & & = 1 \\
& & & & x_4 +\tfrac{1}{2}x_5 & +6s & & = \tfrac{5}{2} \\
& x_1, & x_2, & x_3, & x_4, \quad x_5, s & \geq & 0 \text{ and integer.}
\end{array}$$

Now the new optimal linear programming solution $x = (2, \tfrac{1}{2}, 1, \tfrac{5}{2}, 0)$ is still not integer, as the original variable x_2, and the slack variable x_4 are fractional. The Gomory fractional cut on row 2, in which x_2 is basic, is $\tfrac{1}{2}x_5 \geq \tfrac{1}{2}$ or $-\tfrac{1}{2}x_5 + t = -\tfrac{1}{2}$ with $t \geq 0$ and integer. Adding this constraint and reoptimizing, we obtain

$$\begin{array}{rrrrrrrl}
z = \max \big[7 & & & & & -3s & -t & \big] \\
& x_1 & & & & +s & & = 2 \\
& & x_2 & & & +s & -t & = 1 \\
& & & x_3 & & -5s & -2t & = 2 \\
& & & & x_4 & +6s & +t & = 2 \\
& & & & x_5 & & -t & = 1 \\
& x_1, & x_2, & x_3, & x_4, \quad x_5, s, & t & \geq & 0 \text{ and integer.}
\end{array}$$

Now the linear programming solution is integral, and optimal, and thus $(x_1, x_2) = (2, 1)$ solves the original integer program. ∎

It is natural to also look at the cuts in the space of the original variables.

Example 8.10 (cont) Considering the first cut, and substituting for x_3 and x_4 gives:

$$\tfrac{1}{7}(14 - 7x_1 + 2x_2) + \tfrac{2}{7}(3 - x_2) \geq \tfrac{6}{7}$$

or $x_1 \leq 2$.

In Figure 8.2 we can verify that this inequality is valid and cuts off the fractional solution $(\tfrac{20}{7}, 3)$. Similarly, substituting for x_5 in the second cut $\tfrac{1}{2}x_5 \geq \tfrac{1}{2}$ gives the valid inequality $x_1 - x_2 \leq 1$ in the original variables. ∎

To find a general formula that gives us the cut in terms of the original variables, one can show:

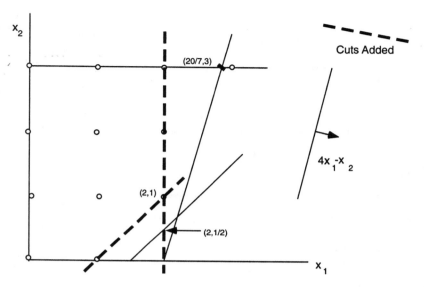

Fig. 8.2 Gomory cutting planes

Proposition 8.5 *Let β be row u of B^{-1}, and $q_i = \beta_i - \lfloor \beta_i \rfloor$ for $i = 1, \cdots, m$. The Gomory cut $\sum_{j \in NB} f_{uj} x_j \geq f_{u0}$, when written in terms of the original variables, is the Chvátal-Gomory inequality*

$$\sum_{j=1}^{n} \lfloor qa_j \rfloor x_j \leq \lfloor qb \rfloor.$$

Looking at the first Gomory cut generated in Example 8.10, β is given by the coefficients of the slack variables in row $u = 1$, so $\beta = (\frac{1}{7}, \frac{2}{7}, 0)$. Thus $q = (\frac{1}{7}, \frac{2}{7}, 0)$ and we obtain $1x_1 + 0x_2 \leq \lfloor \frac{20}{7} \rfloor = 2$.

8.7 MIXED INTEGER CUTS

8.7.1 The Basic Mixed Integer Inequality

We saw above that when $y \leq b, y \in Z^1$, the rounding inequality $y \leq \lfloor b \rfloor$ suffices to generate all the inequalities for a pure integer program. Here we examine if there is a similar *basic* inequality for mixed integer programs.

Proposition 8.6 *Let $X^{\geq} = \{(x, y) \in R^1_+ \times Z^1 : x + y \geq b\}$, and $f = b - \lfloor b \rfloor > 0$. The inequality*

$$x \geq f(\lceil b \rceil - y) \text{ or } \frac{x}{f} + y \geq \lceil b \rceil$$

is valid for $X^{\geq}$.

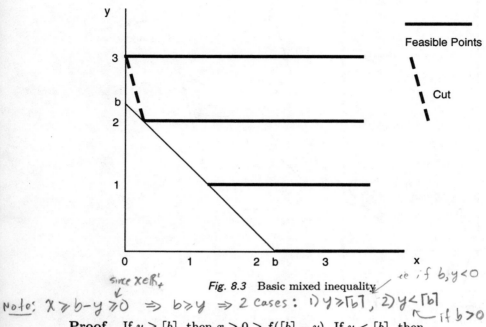

Fig. 8.3 Basic mixed inequality

Note: $x \geq b-y \geq 0 \Rightarrow b \geq y \Rightarrow$ 2 cases: 1) $y \geq \lceil b \rceil$, 2) $y < \lceil b \rceil$
— if $b > 0$
ie if $b, y < 0$
since $x \in \mathbb{R}^1_+$

Proof. If $y \geq \lceil b \rceil$, then $x \geq 0 \geq f(\lceil b \rceil - y)$. If $y < \lceil b \rceil$, then

$$\begin{aligned} x &\geq b - y = f + (\lfloor b \rfloor - y) \\ &\geq f + f(\lfloor b \rfloor - y), \text{ as } \lfloor b \rfloor - y \geq 0 \text{ and } f < 1, \\ &= f(\lceil b \rceil - y). \end{aligned}$$

∎

The situation is shown in Figure 8.3. The following corollary allows us to compare more directly with the all-integer case.

Corollary If $X^{\leq} = \{(x,y) \in R^1_+ \times Z^1 : y \leq b + x\}$, and $f = b - \lfloor b \rfloor > 0$, the inequality

$$y \leq \lfloor b \rfloor + \frac{x}{1-f}$$

is valid for $X^{\leq}$.

Proof. Rewriting $y \leq b + x$ as $x - y \geq -b$ and observing that $-b - \lfloor -b \rfloor = 1 - f$, we obtain from Proposition 8.6 that $\frac{x}{1-f} - y \geq \lceil -b \rceil = -\lfloor b \rfloor$. ∎

Thus we see that when the continuous variable $x = 0$, we obtain the basic integer rounding inequality.

Example 8.6 (cont) The trucking example discussed earlier led to the set $2y_1 + 2y_2 + y_3 + y_4 + \frac{s}{11} \geq \frac{72}{11}$ with $y \in Z^4_+$ and $s \geq 0$. Using Proposition 8.6

with $\lceil b \rceil = 7$ and $f = \frac{6}{11}$, we obtain immediately that

$$\frac{s}{11} \geq \frac{6}{11}(7 - 2y_1 - 2y_2 - y_3 - y_4)$$

is a valid inequality.

8.7.2 The Mixed Integer Rounding (MIR) Inequality

To obtain a slight variant of the basic inequality, we consider a set

$$X^{MIR} = \{(x, y) \in R_+^1 \times Z_+^2 : a_1 y_1 + a_2 y_2 \leq b + x\},$$

where a_1, a_2 and b are scalars with $b \notin Z^1$.

Proposition 8.7 Let $f = b - \lfloor b \rfloor$ and $f_i = a_i - \lfloor a_i \rfloor$ for $i = 1, 2$. Suppose $f_1 \leq f \leq f_2$, then

$$\lfloor a_1 \rfloor y_1 + (\lfloor a_2 \rfloor + \frac{f_2 - f}{1 - f}) y_2 \leq \lfloor b \rfloor + \frac{x}{1 - f} \qquad (8.12)$$

is valid for X^{MIR}.

Proof. $(x, y) \in X^{MIR}$ satisfies $\lfloor a_1 \rfloor y_1 + \lceil a_2 \rceil y_2 \leq b + x + (1 - f_2) y_2$ as $y_1 \geq 0$, and $a_2 = \lceil a_2 \rceil - (1 - f_2)$. Now the Corollary to Proposition 8.6 gives

$$\lfloor a_1 \rfloor y_1 + \lceil a_2 \rceil y_2 \leq \lfloor b \rfloor + [x + (1 - f_2) y_2]/(1 - f),$$

which is the required inequality. ∎

Example 8.11 Consider the set $X = \{(y, x) \in Z_+^3 \times R_+^1 : \frac{10}{3} y_1 + 1 y_2 + \frac{11}{4} y_3 \leq \frac{21}{2} + x\}$. Using Proposition 8.7, we have $f = 1/2, f_1 = 1/3, f_2 = 0, f_3 = 3/4$, and thus

$$3y_1 + y_2 + \frac{5}{2} y_3 \leq 10 + 2x$$

is valid for X. ∎

8.7.3 The Gomory Mixed Integer Cut*

Here we continue to consider mixed integer programs. As for all integer programs in Section 8.6, any row of the optimal linear programming tableau, in which an integer variable is basic but fractional, can be used to generate a cut removing the optimal linear programming solution. Specifically, such a row leads to a set of the form:

$$X^G = \{(y_{B_u}, y, x) \in Z^1 \times Z_+^{n_1} \times R_+^{n_2} : y_{B_u} + \sum_{j \in N_1} \bar{a}_{uj} y_j + \sum_{j \in N_2} \bar{a}_{uj} x_j = \bar{a}_{u0}\},$$

where $n_i = |N_i|$ for $i = 1, 2$.

Proposition 8.8 *If $\bar{a}_{u0} \notin Z^1$, $f_j = \bar{a}_{uj} - \lfloor \bar{a}_{uj} \rfloor$ for $j \in N_1 \cup N_2$, and $f_0 = \bar{a}_{u0} - \lfloor \bar{a}_{u0} \rfloor$, the Gomory mixed integer cut*

$$\sum_{f_j \leq f_0} f_j y_j + \sum_{f_j > f_0} \frac{f_0(1-f_j)}{1-f_0} y_j + \sum_{\bar{a}_{uj} > 0} \bar{a}_{uj} x_j + \sum_{\bar{a}_{uj} < 0} \frac{f_0}{1-f_0} \bar{a}_{uj} x_j \geq f_0$$

is valid for X^G.

Proof. The mixed integer rounding inequality (8.12) for X^G is

$$y_{B_u} + \sum_{f_j \leq f_0} \lfloor \bar{a}_{uj} \rfloor y_j + \sum_{f_j > f_0} (\lfloor \bar{a}_{uj} \rfloor + \frac{f_j - f_0}{1 - f_0}) y_j + \sum_{\bar{a}_{uj} < 0} \frac{\bar{a}_{uj}}{1-f_0} x_j \leq \lfloor \bar{a}_{u0} \rfloor.$$

Substituting for y_{B_u} proves the claim. ∎

Example 8.12 Consider the mixed integer program:

$$\begin{array}{rrrcl}
z = \max & 4x_1 & - x_2 & & \\
& 7x_1 & - 2x_2 & \leq & 14 \\
& & x_2 & \leq & 3 \\
& 2x_1 & - 2x_2 & \leq & 3 \\
& x_1 & \in Z_+^1, \ x_2 & \geq & 0.
\end{array}$$

Note that this is the same as Example 8.10 except that variable $x_2 \in R_+^1$ is a real variable. Solving as a linear program gives:

$$\begin{array}{rrrrcl}
z = \max \frac{59}{7} & & -\frac{4}{7}x_3 & -\frac{1}{7}x_4 & & \\
& x_1 & +\frac{1}{7}x_3 & +\frac{2}{7}x_4 & = & \frac{20}{7} \\
& x_2 & & +x_4 & = & 3 \\
& & -\frac{2}{7}x_3 & +\frac{10}{7}x_4 & +x_5 = & \frac{23}{7} \\
& x_1 \in Z_+^1, & x_2, & x_3, & x_4, x_5 \geq & 0.
\end{array}$$

The basic variable x_1 is fractional and the first row gives the MIR cut $x_1 \leq 2$, which after elimination of x_1 becomes the Gomory mixed integer cut:

$$\tfrac{1}{7}x_3 + \tfrac{2}{7}x_4 \geq \tfrac{6}{7}.$$

Adding this cut and reoptimizing leads to the solution $x = (2, \tfrac{1}{2})$, which is feasible and hence optimal for the mixed integer program. This can also be seen graphically in Figure 8.2 with just the addition of the cut $x_1 \leq 2$. ∎

8.8 DISJUNCTIVE INEQUALITIES*

The set $X = X^1 \cup X^2$ with $X^i \subseteq R_+^n$ for $i = 1, 2$ is a *disjunction* (union) of the two sets X^1 and X^2. The following simple result has already been used implicitly in Proposition 8.6 in deriving the basic mixed integer inequality.

Proposition 8.9 *If $\sum_{j=1}^{n} \pi_j^i x_j \leq \pi_0^i$ is valid for X^i for $i = 1, 2$, then the inequality*

$$\sum_{j=1}^{n} \pi_j x_j \leq \pi_0$$

is valid for X if $\pi_j \leq \min[\pi_j^1, \pi_j^2]$ for $j = 1, \ldots, n$ and $\pi_0 \geq \max[\pi_0^1, \pi_0^2]$.

Proof. If $x \in X$, then $x \in X^1$ or $x \in X^2$. If $x \in X^i$, then as $x \geq 0$, $\sum_{j=1}^{n} \pi_j x_j \leq \sum_{j=1}^{n} \pi_j^i x_j \leq \pi_0^i \leq \pi_0$ for $i = 1, 2$. Thus the inequality is valid for all $x \in X$. ∎

Disjunctions of polyhedra are particularly interesting. Modeling such sets is easy; see Exercises 1.3 and 8.10. Using Proposition 8.2, it is also easy to characterize valid inequalities for such disjunctions.

Proposition 8.10 *If $P^i = \{x \in R^n_+ : A^i x \leq b^i\}$ for $i = 1, 2$ are nonempty polyhedra, then (π, π_0) is a valid inequality for $\text{conv}(P^1 \cup P^2)$ if and only if there exist $u^1, u^2 \geq 0$ such that $\pi \leq u^i A^i$ and $\pi_0 \geq u^i b^i$ for $i = 1, 2$.*

Example 8.13 Let $P^1 = \{x \in R^2_+ : -x_1 + x_2 \leq 1, x_1 + x_2 \leq 5\}$ and $P^2 = \{x \in R^2_+ : x_2 \leq 4, -2x_1 + x_2 \leq -6, x_1 - 3x_2 \leq -2\}$. Taking $u^1 = (2, 1)$ and $u^2 = (\frac{5}{2}, \frac{1}{2}, 0)$ and applying Proposition 8.10 gives that $-x_1 + 3x_2 \leq 7$ is valid for $P^1 \cup P^2$. See Figure 8.4. ∎

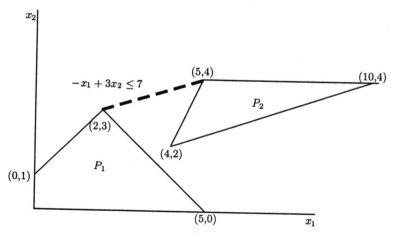

Fig. 8.4 Disjunctive inequality

Specializing further, we consider 0–1 problems, where $X = P \cap Z^n \subseteq \{0, 1\}^n$ with $P = \{x \in R^n : Ax \leq b, \mathbf{0} \leq x \leq \mathbf{1}\}$. Let $P^0 = P \cap \{x \in R^n : x_j = 0\}$, and $P^1 = P \cap \{x \in R^n : x_j = 1\}$ for some $j \in \{1, \ldots, n\}$.

Proposition 8.11 *The inequality* (π, π_0) *is a valid for* $conv(P^0 \cup P^1)$ *if there exists* $u^i \in R_+^m, v^i \in R_+^n, w^i \in R_+^1$ *for* $i = 0, 1$ *such that*

$$\pi \leq u^0 A + v^0 + w^0 e_j, \pi \leq u^1 A + v^1 - w^1 e_j,$$
$$\pi_0 \geq u^0 b + 1.v^0, \pi_0 \geq u^1 b + 1.v^1 - w^1.$$

Proof. Apply Proposition 8.10 with $P^0 = \{x \in R_+^n : Ax \leq b, x \leq 1, x_j \leq 0\}$ and $P^1 = \{x \in R_+^n : Ax \leq b, x \leq 1, -x_j \leq -1\}$. ∎

Example 8.14 Consider the 0–1 knapsack problem

$$\max 12x_1 + 14x_2 + 7x_3 + 12x_4$$
$$4x_1 + 5x_2 + 3x_3 + 6x_4 \leq 8$$
$$x \in B^4$$

with linear programming solution $x^* = (1, 0.8, 0, 0)$.

As $x_2^* = 0.8$ is fractional, we choose $j = 2$ in defining P^0 and P^1, and then look for the most violated valid inequality (π, π_0) given by Proposition 8.11. To do this, we solve a linear program consisting of maximizing $\pi x^* - \pi_0$ over the polyhedron describing the coefficients of the valid inequalities given in the proposition, namely

$$\max 1.0\pi_1 + 0.8\pi_2 - \pi_0$$
$$\pi_1 \leq 4u^0 + v_1^0, \pi_1 \leq 4u^1 + v_1^1$$
$$\pi_2 \leq 5u^0 + v_2^0 + w^0, \pi_2 \leq 5u^1 + v_2^1 - w^1$$
$$\pi_3 \leq 3u^0 + v_3^0, \pi_3 \leq 3u^1 + v_3^1$$
$$\pi_4 \leq 6u^0 + v_4^0, \pi_4 \leq 6u^1 + v_4^1$$
$$\pi_0 \geq 8u^0 + v_1^0 + v_2^0 + v_3^0 + v_4^0$$
$$\pi_0 \geq 8u^1 + v_1^1 + v_2^1 + v_3^1 + v_4^1 - w^1$$
$$u^0, u^1, v^0, v^1, w^0, w^1 \geq 0.$$

Note that for the linear program to have a bounded optimal value, it is necessary to normalize the inequality. Two possibilities are $\sum_{j=1}^n \pi_j \leq 1$ or $\pi_0 = 1$. The resulting inequality is

$$1x_1 + \tfrac{1}{4}x_2 \leq 1,$$

with violation of $\frac{1}{5}$. For P^0, it is a combination of constraints $x_1 \leq 1$ and $x_2 \leq 0$ with $v_1^0 = 1$ and $w^0 = \frac{1}{4}$ respectively. For P^1, it is a combination of the knapsack inequality $4x_1 + 5x_2 + 3x_3 + 6x_4 \leq 8$ and $-x_2 \leq -1$ with $u^1 = \frac{1}{4}$ and $w^1 = 1$ respectively. Both normalizations lead to the same inequality. ∎

The idea of looking for the most violated inequality will be pursued in the next chapter, when we try to obtain "strong" inequalities.

8.9 NOTES

8.2 The inequality (8.4), called a *blossom inequality*, is from [Edm65a].

8.3 The rounding procedure to generate cuts is from [Gom58]. The general procedure described here and the proof of Theorem 8.4 for bounded integer programs is from [Chv73]. In [Sch80], the result is extended to unbounded polyhedra.

8.4 The strong formulation for UFL is used computationally in [Spi69].

8.5 The first cutting plane algorithm reported is the procedure used to solve a 54-city TSP in [DanFulJoh54].

8.6 The fractional cutting plane algorithm is presented in [Gom58],[Gom63]. The latter paper also contains a beautiful theoretical result, namely that the algorithm converges finitely if the rows off which the cuts are generated are properly chosen.

8.7 Gomory mixed integer cuts are proposed in [Gom60]. The presentation of mixed integer rounding inequalities is from [NemWol90]. The theory of superadditive valid inequalities and superadditive duality [Joh80] and Chapter II.1 in [NemWol88] provides a complete explanation of cuts for integer and mixed integer programs.

Gomory has shown that finite convergence can be attained with his mixed integer cuts if the objective function is integer valued. It is an open question whether this is true for 0–1 mixed integer programs with an arbitrary objective function. In [CooKanSch90], the question of how finite convergence might be obtained is reexamined.

Recently [GunPoc98] present a new way to combine basic mixed integer inequalities. Gomory mixed integer cuts have also been recently revived as a computational tool; see [Balasetal96].

8.8 Disjunctive and Gomory mixed integer cuts are closely related. Proposition 8.9 was already used implicitly by Gomory in developing the mixed integer cut. In the same way that the Chvátal-Gomory procedure can be used to generate all valid inequalities for an integer program, it can be shown that a simple disjunctive procedure repeated finitely (see Exercise 8.11) can be used to generate all valid inequalities for a 0–1 mixed integer program.

The approach here is based on the disjunction of polyhedra developed by Balas [Balas75a] in the 1970s; see also [Jer72]. In particular, Balas shows the beautiful result that to obtain the convex hull of a 0–1 MIP, it suffices to take the convex hulls of each 0–1 variable one at a time. Related to this result, a variety of extended formulations have recently been proposed to obtain tighter formulations for 0–1 $MIPs$ [LovSch91], [BalasCerCor93], [SheAda90].

8.10 EXERCISES

1. For each of the three sets below, find a missing valid inequality and verify graphically that its addition to the formulation gives conv(X).

(i) $X = \{x \in B^2 : 3x_1 - 4x_2 \leq 1\}$
(ii) $X = \{(x,y) \in R^1_+ \times B^1 : x \leq 20y, x \leq 7\}$
(iii) $X = \{(x,y) \in R^1_+ \times Z^1_+ : x \leq 6y, x \leq 16\}$

2. In each of the examples below a set X and a point x or (x,y) are given. Find a valid inequality for X cutting off the point.

(i)
$$X = \{(x,y) \in R^2_+ \times B^1 : x_1 + x_2 \leq 2y, x_j \leq 1 \text{ for } j = 1,2\}$$
$$(x_1, x_2, y) = (1, 0, 0.5)$$

(ii)
$$X = \{(x,y) \in R^1_+ \times Z^1_+ : x \leq 9, x \leq 4y\}$$
$$(x,y) = (9, \frac{9}{4})$$

(iii)
$$X = \{(x_1, x_2, y) \in R^2_+ \times Z^1_+ : x_1 + x_2 \leq 25, x_1 + x_2 \leq 8y\}$$
$$(x_1, x_2, y) = (20, 5, \frac{25}{8})$$

(iv)
$$X = \{x \in Z^5_+ : 9x_1 + 12x_2 + 8x_3 + 17x_4 + 13x_5 \geq 50\}$$
$$x = (0, \frac{25}{6}, 0, 0, 0)$$

(v)
$$X = \{x \in Z^4_+ : 4x_1 + 8x_2 + 7x_3 + 5x_4 \leq 33\}$$
$$x = (0, 0, \frac{33}{7}, 0).$$

3. Prove that $y_2 + y_3 + 2y_4 \leq 6$ is valid for
$$X = \{y \in Z^4_+ : 4y_1 + 5y_2 + 9y_3 + 12y_4 \leq 34\}.$$

4. Consider the problem
$$\min x_1 + 2x_2$$
$$x_1 + x_2 \geq 4$$
$$\tfrac{1}{2}x_1 + \tfrac{5}{2}x_2 \geq \tfrac{5}{2}$$
$$x \in Z_+^2.$$

Show that $x^* = (\tfrac{15}{4}, \tfrac{1}{4})$ is the optimal linear programming solution and find an inequality cutting off x^*.

5. Solve $\min\{5x_1 + 9x_2 + 23x_3 : 20x_1 + 35x_2 + 95x_3 \geq 319, x \in Z_+^3\}$ using Chvátal-Gomory inequalities or Gomory's cutting plane algorithm.

6. Solve $\max\{5x_1 + 9x_2 + 23x_3 - 4s : 2x_1 + 3x_2 + 9x_3 \leq 32 + s, x \in Z_+^3, s \in R_+^1\}$ using MIR inequalities.

7. (i) Show that the inequality $x_t \leq d_t y_t + s_t$ is valid for ULS.
(ii) Show that $x_t + x_{t+1} \leq (d_t + d_{t+1})y_t + d_{t+1}y_{t+1} + s_{t+1}$ is valid. (Hint: use cases for y_t, y_{t+1})
(iii) For $l \leq n$, $L = \{1, \ldots, l\}$ and $S \subseteq L$, show that the inequality

$$\sum_{j \in S} x_j \leq \sum_{j \in S}(\sum_{t=j}^{l} d_t)y_j + s_l$$

is valid for ULS.

8. Consider the stable set problem. An *odd hole* is a cycle with an odd number of nodes and no edges between nonadjacent nodes of the cycle. Show that if H is the node set of an odd hole,

$$\sum_{j \in H} x_j \leq (|H| - 1)/2$$

is a valid inequality.

9. Use the mixed integer rounding procedure to show that

$$(y_1 + 6y_2)/4 + y_3 + 4y_4 \geq 16$$

is a valid inequality for

$$X = \{y \in Z_+^4 : y_1 + 6y_2 + 12y_3 + 48y_4 \geq 184\}.$$

10. Use the mixed integer rounding procedure to show that

$$x_2 + x_4 \leq 20 + 4(y - 2)$$

is a valid inequality for $X =$

$\{(x, y) \in R_+^4 \times Z_+^1 : x_1 + x_2 + x_3 + x_4 \leq 10y, x_1 \leq 13, x_2 \leq 15, x_3 \leq 6, x_4 \leq 9\}.$

11.(i) Show that if $\pi x \leq \pi_0 + \alpha(x_j - k)$ and $\pi x \leq \pi_0 + \beta(k+1-x_j)$ with $\alpha, \beta > 0$ and $k \in Z^1$ are both valid for a polyhedron P, then $\pi x \leq \pi_0$ is valid for $P \cap \{x : x_j \in Z^1\}$. An inequality generated in this way is called a *D-inequality*.

(ii)* Show that if $P \subseteq R^n$ is a polyhedron and $j \in \{1, \ldots, n\}$, every valid inequality for $conv(P \cap \{x : x_j \in B^1\})$ is or is dominated by a D-inequality.

12.* Prove that if $P^k = \{x \in R^n : A^k x \leq b^k\}$ are bounded polyhedra for $k = 1, 2$, then $conv(P^1 \cup P^2) = \{x : \text{there exists } (x, z^1, z^2, y^1, y^2) \in R^n \times R^n \times R^n \times R^1_+ \times R^1_+ \text{ satisfying } A^k z^k \leq b^k y^k \text{ for } k = 1, 2, x = z^1 + z^2, y^1 + y^2 = 1\}$.

13. Consider an instance of the generalized transportation problem

$$\min \sum_{i=1}^m \sum_{j=1}^n c_{ij} x_{ij}$$
$$\sum_{j=1}^n x_{ij} \leq b_i \text{ for } i = 1, \ldots, m$$
$$\sum_{i=1}^m a_i x_{ij} \geq d_j \text{ for } j = 1, \ldots, n$$
$$x \in Z_+^{mn}$$

with $m = 4, n = 6, a = (15, 25, 40, 70), b = (10, 5, 7, 4), d = (45, 120, 165, 214, 64, 93)$ and

$$(c_{ij}) = \begin{pmatrix} 23 & 12 & 34 & 25 & 27 & 16 \\ 29 & 24 & 43 & 35 & 28 & 19 \\ 43 & 31 & 52 & 36 & 30 & 21 \\ 54 & 36 & 54 & 46 & 34 & 27 \end{pmatrix}.$$

Solve with a mixed integer programming system. Now by inspecting the linear programming solution, or otherwise, add one or more valid inequalities to the formulation and resolve. Compare the number of nodes in the branch-and-bound tree before and after. Try to minimize the number of nodes.

14. Consider a telecommunications problem where the demands between pairs of nodes are given. The problem is to install sufficient capacity on the edges of the graph so that all the demands can be satisfied simultaneously. If there is flow of one demand type from i to j, and simultaneously others from j to i, the capacity available must be the sum of the two opposite flows. Capacity can be installed in units of 1 and/or 24, costing 1 and 10 respectively. For a graph on 6 nodes, the following demand matrix must be satisfied

$$(d_{ij}) = \begin{pmatrix} . & 12 & 51 & - & - & . \\ . & . & 53 & 51 & - & . \\ . & . & . & - & 32 & . \\ . & . & . & . & 91 & . \\ . & . & . & . & . & . \end{pmatrix}$$

Formulate and solve with a mixed integer programming system. Try to tighten the formulation.

15. (i) Derive the inequalities of Example 8.1 as C-G inequalities.
(ii) Consider the set $X = \{x \in B^4 : x_i + x_j \leq 1 \text{ for all } 1 \leq i < j \leq 4\}$. Derive the clique inequalities $x_1 + x_2 + x_3 \leq 1$ and $x_1 + x_2 + x_3 + x_4 \leq 1$ as C-G inequalities.

9
Strong Valid Inequalities

9.1 INTRODUCTION

In the last chapter we have seen a variety of valid inequalities, and presented a generic cutting plane algorithm. The Gomory fractional cutting plane algorithm is a special case of this algorithm with the particularity that finding cuts is very easy at each iteration. Theoretically it is of interest because it has been shown to terminate after a finite number of iterations, but in practice it has not been successful. However, Gomory mixed integer cuts, as well as the disjunctive cuts, have been recently successfully used in practice.

Here we address the question of finding *strong* valid inequalities that are hopefully even more effective. The basic cutting plane algorithm is the same as in Section 8.5, however:

(i) We need to say what "strong" means — for our purposes it is any inequality that leads to a stronger formulation. However, in Section 9.2 (optional) we formalize what is meant by the strength of an inequality.

(ii) Describing interesting families $\mathcal{F}$ of strong inequalities may be far from easy.

(iii) Given a family $\mathcal{F}$ of strong valid inequalities, the separation problem for $\mathcal{F}$ may require a lot of work. It may be polynomially solvable, or it may be $\mathcal{NP}$-hard, in which case a heuristic algorithm has to be developed. In Sections 9.3–9.5, we examine three sets: 0–1 knapsack sets, mixed 0–1 sets, and the set of incidence vectors of subtours that arises in a generalization of the

140   STRONG VALID INEQUALITIES

traveling salesman problem. We develop a family of strong valid inequalities, and discuss the resulting separation problem for each of the families.

(iv) To solve difficult or large problems to optimality, strong cutting planes need to be embedded into a branch-and-bound framework. The resulting algorithms, called branch-and-cut algorithms, are discussed in Section 9.6.

In discussing separation algorithms, it is important to remember certain ideas encountered earlier. First because Efficient (Polynomial) Optimization and Efficient Separation are equivalent,

(i) If Efficient Optimization holds for a class of problems of the form $\max\{cx : x \in X\}$, it may be possible to obtain an "explicit" description of the convex hull of X, and perhaps also a combinatorial separation algorithm for $\text{conv}(X)$, and

(ii) If the Optimization Problem is $\mathcal{NP}$-hard, there is no hope (unless $\mathcal{P} = \mathcal{NP}$) of obtaining an explicit description of $\text{conv}(X)$, but this should not deter us from looking for families of strong valid inequalities.

As before, the idea of *decomposition* may be useful. In particular,

(iii) If we can break up the feasible set so that $X = X^1 \cap X^2$ where the optimization problem over X^2 is polynomially solvable, then we can attempt to find $\text{conv}(X^2)$, and

(iv) If $X = X^1 \cap X^2$, but the optimization problems over X^1 and X^2 are both $\mathcal{NP}$-hard, it may be still be worthwhile (and easier) to attempt to find valid inequalities for X^1 and X^2 separately in the hope that the resulting inequalities will also be strong for the intersection $X = X^1 \cap X^2$.

Note: Since $X^1 \cap X^2 \subseteq X^2$, any valid inequality for X^2 is also valid for $X^1 \cap X^2$ (similarly for X^1)

9.2   STRONG INEQUALITIES

Here we address briefly the question of what it means for an inequality to be strong for a set $P = \{x \in R_+^n : Ax \leq b\}$. This leads us to introduce certain concepts important for the description of polyhedra. We also present different arguments that can be used to prove that an inequality is strong, or to show that a set of inequalities describes the convex hull of a discrete set.

9.2.1   Dominance

We note first that the inequalities $\pi x \leq \pi_0$ and $\lambda \pi x \leq \lambda \pi_0$ are identical for any $\lambda > 0$.

> e.g. $2x_1 + 6x_2 \leq 8$ dominates $2x_1 + 4x_2 \leq 9$
> $\pi = (2,6)$, $\mu = (2,4)$ and $\pi \geq \mu$ $(u=1)$
> $\pi_0 = 8$, $\mu_0 = 9$ and $\pi_0 \leq \mu_0$

STRONG INEQUALITIES 141

Definition 9.1 If $\pi x \leq \pi_0$ and $\mu x \leq \mu_0$ are two valid inequalities for $P \subseteq R_+^n$, $\pi x \leq \pi_0$ *dominates* $\mu x \leq \mu_0$ if there exists $u > 0$ such that $\pi \geq u\mu$ and $\pi_0 \leq u\mu_0$, and $(\pi, \pi_0) \neq (u\mu, u\mu_0)$.

Observe that if $\pi x \leq \pi_0$ dominates $\mu x \leq \mu_0$, then $\{x \in R_+^n : \pi x \leq \pi_0\} \subseteq \{x \in R_+^n : \mu x \leq \mu_0\}$.

Definition 9.2 A valid inequality $\pi x \leq \pi_0$ is *redundant* in the description of P, if there exist $k \geq 2$ valid inequalities $\pi^i x \leq \pi_0^i$ for $i = 1, \ldots, k$ for P, and weights $u_i > 0$ for $i = 1, \ldots, k$ such that $(\sum_{i=1}^k u_i \pi^i) x \leq \sum_{i=1}^k u_i \pi_0^i$ dominates $\pi x \leq \pi_0$.

Here we observe that $\{x \in R_+^n : \pi^i x \leq \pi_0^i \text{ for } i = 1, \ldots, k\} \subseteq \{x \in R_+^n : (\sum_{i=1}^k u_i \pi^i) x \leq \sum_{i=1}^k u_i \pi_0^i\} \subseteq \{x \in R_+^n : \pi x \leq \pi_0\}$.

Example 9.1 Taking $n = 2$, $(\pi, \pi_0) = (1, 3, 4)$ and $(\mu, \mu_0) = (2, 4, 9)$, we see that with $u = \frac{1}{2}$, $\pi \geq \frac{1}{2}\mu$ and $\pi_0 \leq \frac{1}{2}\mu_0$, and so $x_1 + 3x_2 \leq 4$ dominates $2x_1 + 4x_2 \leq 9$. See Figure 9.1a.

$\Rightarrow 2x_1 + 6x_2 \leq 8$

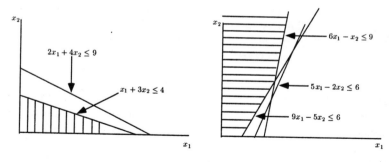

Fig. 9.1 Dominance of inequalities

(Redundant is like whom multiple inequalities combined "dominate" another)

Again with $n = 2$, suppose that $P = \{x \in R_+^2 : 6x_1 - x_2 \leq 9, 9x_1 - 5x_2 \leq 6\}$. Now consider another valid inequality $5x_1 - 2x_2 \leq 6$ for P. Taking weights $u = (\frac{1}{3}, \frac{1}{3})$, we see that $5x_1 - 2x_2 \leq 6$ is redundant. See Figure 9.1b. ∎

Where $P = \text{conv}(X)$ is not known explicitly, checking redundancy may be very difficult. Theoretically it is important to know which inequalities are needed or nonredundant in the description of P. Practically, the important point is to avoid using an inequality when one that dominates it is readily available.

In the next subsection, we discuss polyhedra and characterize which inequalities are nonredundant.

9.2.2 Polyhedra, Faces, and Facets

The goal here is to understand which are the important inequalities that are necessary in describing a polyhedon, and hence at least in theory provide the best possible cuts.

For simplicity we limit the discussion to polyhedra $P \subseteq R^n$ that contain n linearly independent directions. Such polyhedra are called *full-dimensional*. Full-dimensional polyhedra have the property that there is no equation $ax = b$ satisfied at equality by all points $x \in P$.

Theorem 9.1 *If P is a full-dimensional polyhedron, it has a unique minimal description*

$$P = \{x \in R^n : a^i x \leq b_i \text{ for } i = 1, \ldots, m\},$$

where each inequality is unique to within a positive multiple.

This means that if one of the inequalities in the minimal description is removed, the resulting polyhedron is no longer P, so each of the inequalities is *necessary*. On the other hand every valid inequality $\pi x \leq \pi_0$ for P that is not a positive multiple of one of the inequalities $a^i x \leq b_i$ for some i with $1 \leq i \leq m$ is redundant in the sense of Definition 9.2 as it is a nonnegative combination of two or more valid inequalities.

We now discuss another way in which the necessary inequalities can be characterized.

Definition 9.3 The points $x^1, \ldots, x^k \in R^n$ are *affinely independent* if the $k-1$ directions $x^2 - x^1, \ldots, x^k - x^1$ are linearly independent, or alternatively the k vectors $(x^1, 1) \ldots, (x^k, 1) \in R^{n+1}$ are linearly independent.

Definition 9.4 The *dimension* of P, denoted $\dim(P)$, is one less than the maximum number of affinely independent points in P.

This means that $P \subseteq R^n$ is full-dimensional if and only if $\dim(P) = n$.

Definition 9.5 (i) F defines a *face* of the polyhedron P if $F = \{x \in P : \pi x = \pi_0\}$ for some valid inequality $\pi x \leq \pi_0$ of P.
(ii) F is a *facet* of P if F is a face of P and $\dim(F) = \dim(P) - 1$.
(iii) If F is a face of P with $F = \{x \in P : \pi x = \pi_0\}$, the valid inequality $\pi x \leq \pi_0$ is said to *represent* or *define* the face.

It follows that the faces of polyhedra are polyhedra, and it can be shown that the number of faces of a polyhedron is finite. Now we establish a way to recognize the necessary inequalities.

Proposition 9.2 *If P is full-dimensional, a valid inequality $\pi x \leq \pi_0$ is necessary in the description of P if and only if it defines a facet of P.*

STRONG INEQUALITIES

So for full-dimensional polyhedra, $\pi x \leq \pi_0$ defines a facet of P if and only if there are n affinely independent points of P satisfying it at equality.

Example 9.2 Consider the polyhedron $P \subset R^2$, shown in Figure 9.2, described by the inequalities

$$\begin{array}{rcl} x_1 & \leq & 2 \\ x_1 + x_2 & \leq & 4 \\ x_1 + 2x_2 & \leq & 10 \\ x_1 + 2x_2 & \leq & 6 \\ x_1 + x_2 & \geq & 2 \\ x_1 & \geq & 0 \\ x_2 & \geq & 0. \end{array}$$

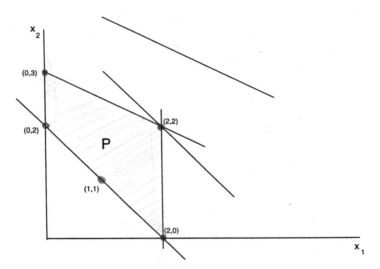

Fig. 9.2 Facets and faces of a polyhedron

P is full-dimensional as $(2,0), (1,1)$, and $(2,2)$ are three affinely independent points in P.

The inequality $x_1 \leq 2$ defines a facet of P as $(2,0)$ and $(2,2)$ are two affinely independent points in P satisfying $x_1 \leq 2$ at equality. Similarly the inequalities $x_1 + 2x_2 \leq 6, x_1 + x_2 \geq 2$ and $x_1 \geq 0$ define facets.

On the other hand, the inequality $x_1 + x_2 \leq 4$ defines a face consisting of just one point $(2,2)$ of P, and hence it is redundant. Alternatively, considering the inequalities $x_1 \leq 2$ and $x_1 + 2x_2 \leq 6$ with weights $u = (\frac{1}{2}, \frac{1}{2})$ also shows that $x_1 + x_2 \leq 4$ is redundant.

The inequality $x_2 \geq 0$ is the sum of the inequalities $x_1 \leq 2$ and $-x_1 - x_2 \leq -2$ and so it is also redundant.

144   STRONG VALID INEQUALITIES

The minimal description is given by

$$\begin{aligned} x_1 &\leq 2 \\ 2x_1 + x_2 &\leq 6 \\ x_1 + x_2 &\geq 2 \\ x_1 &\geq 0. \end{aligned}$$

9.2.3 Facet and Convex Hull Proofs*

This section is for those interested in proving results about the strength of certain inequalities or formulations. The aim is to indicate ways to show that a valid inequality is facet-defining, or that a set of inequalities describes the convex hull of some discrete set $X \subset Z_+^n$.

For simplicity we assume throughout this subsection that $\mathrm{conv}(X)$ is bounded as well as full-dimensional. So there are no hyperplanes containing all the points of X. As example we take the set $X = \{(x,y) \in R_+^m \times B^1 : \sum_{i=1}^m x_i \leq my\}$ that arises in Sections 1.6 and 8.4 in formulating the uncapacitated facility location problem.

Problem 1. Given $X \subset Z_+^n$ and a valid inequality $\pi x \leq \pi_0$ for X, show that the inequality defines a facet of $\mathrm{conv}(X)$.

We consider two different approaches.

Approach 1. (Just use the definition.) Find n points $x^1, \ldots, x^n \in X$ satisfying $\pi x = \pi_0$, and then prove that these n points are affinely independent.
(in general, find dim(P) # affinely independent points)

Approach 2. (An indirect but useful way to verify the affine independence.)
(i) Select $t \geq n$ points $x^1, \ldots, x^t \in X$ satisfying $\pi x = \pi_0$. Suppose that all these points lie on a generic hyperplane $\mu x = \mu_0$.
(ii) Solve the linear equation system

$$\sum_{j=1}^n \mu_j x_j^k = \mu_0 \text{ for } k = 1, \ldots, t$$

in the $n+1$ unknowns (μ, μ_0).
(iii) If the only solution is $(\mu, \mu_0) = \lambda(\pi, \pi_0)$ for $\lambda \neq 0$, then the inequality $\pi x \leq \pi_0$ is facet-defining.

Too Time Consuming

Example 9.3 Taking $X = \{(x,y) \in R_+^m \times B^1 : \sum_{i=1}^m x_i \leq my\}$, we have that $\dim(\mathrm{conv}(X)) = m+1$. Now we consider the valid inequality $x_i \leq y$ and show that it is facet-defining using Approach 2.

We select the simplest points $(0,0), (e_i, 1)$ and $(e_i + e_j, 1)$ for $j \neq i$ that are feasible and satisfy $x_i = y$.

STRONG INEQUALITIES   145

As $(0,0)$ lies on $\sum_{i=1}^{m} \mu_i x_i + \mu_{m+1} y = \mu_0$, $\mu_0 = 0$.
As $(e_i, 1)$ lies on the hyperplane $\sum_{i=1}^{m} \mu_i x_i + \mu_{m+1} y = 0$, $\mu_i = -\mu_{m+1}$.
As $(e_i + e_j, 1)$ lies on the hyperplane $\sum_{i=1}^{m} \mu_i x_i - \mu_i y = 0$, $\mu_j = 0$ for $j \neq i$.
So the hyperplane is $\mu_i x_i - \mu_i y = 0$, and $x_i \leq y$ is facet-defining. ∎

Problem 2. Show that the polyhedron $P = \{x \in R^n : Ax \leq b\}$ describes $\mathrm{conv}(X)$. (where $X = P \cap Z^n$)

Here we present eight approaches.

Approach 1. Show that the matrix A, or the pair (A, b) have special structure guaranteeing that $P = \mathrm{conv}(X)$.

Example 9.4 Take $X = \{(x, y) \in R_+^m \times B^1 : \sum_{i=1}^{m} x_i \leq my\}$, and consider the polyhedron/formulation

$$P = \{(x, y) \in R_+^m \times R^1 : x_i \leq y \text{ for } i = 1, \ldots, m, y \leq 1\}.$$

Observe that the constraints $x_i - y \leq 0$ for $i = 1, \ldots, m$ lead to a matrix with a coefficient of $+1$ and -1 in each row. Such a matrix is TU; see Proposition 3.2. Adding the bound constraints still leaves a TU matrix. Now as the requirements vector is integer, it follows from Proposition 3.3 that all basic solutions are integral, and $P = \mathrm{conv}(X)$. ∎

Approach 2. Show that points $(x, y) \in P$ with y fractional are not extreme points of P. (That is, take an arbitrary fractional point and show its not extreme)

Example 9.4 (cont) Suppose that $(x^*, y^*) \in P$ with $0 < y^* < 1$. Note first that $(0, 0) \in P$. Also as $x_i^* \leq y^*$, the point $(\frac{x_1^*}{y^*}, \ldots, \frac{x_m^*}{y^*}, 1) \in P$. But now

$$(x^*, y^*) = (1 - y^*)(0, 0) + y^* (\frac{x_1^*}{y^*}, \ldots, \frac{x_m^*}{y^*}, 1)$$

is a convex combination of two points of P and is not extreme. Thus all vertices of P have y^* integer. ∎

Approach 3. Show that for all $c \in R^n$, the linear program $z^{LP} = \max\{cx : Ax \leq b\}$ has an optimal solution $x^* \in X$.

Example 9.4 (cont) Consider the linear program $z^{LP} = \max\{\sum_{i=1}^{m} c_i x_i + fy : 0 \leq x_i \leq y \text{ for } i = 1, \ldots, m, y \leq 1\}$. Consider an optimal solution (x^*, y^*). Because of the constraints $0 \leq x_i \leq y$, any optimal solution has $x_i^* = y^*$ if $c_i > 0$ and $x_i^* = 0$ if $c_i < 0$. The corresponding solution value is $(\sum_{i:c_i>0} c_i + f) y^*$ if $y^* > 0$ and 0 otherwise. Obviously if $(\sum_{i:c_i>0} c_i + f) > 0$, the objective is maximized by setting $y^* = 1$, and otherwise $y^* = 0$ is optimal. Thus there is always an optimal solution with y integer, and $z^{LP} = (\sum_{i:c_i>0} c_i + f)^+$. ∎

146   STRONG VALID INEQUALITIES

Approach 4. Show that for all $c \in R^n$, there exists a point $x^* \in X$ and a feasible solution u^* of the dual LP $w^{LP} = \min\{ub, uA = c, u \geq 0\}$ with $cx^* = u^*b$. Note that this implies that the condition of Approach 3 is satisfied.

Example 9.4 (cont) The dual linear program is

$$\min t$$
$$w_i \geq c_i \text{ for } i = 1, \ldots, m$$
$$-\sum_{i=1}^m w_i + t \geq f$$
$$w_i \geq 0 \text{ for } i = 1, \ldots, m, t \geq 0.$$

Consider the two points $(\mathbf{0}, 0)$ and $(x^*, 1)$ with $x_i^* = 1$ if $c_i > 0$ and $x_i^* = 0$ otherwise. Taking the better of the two leads to a primal solution of value $(\sum_{i:c_i>0} c_i + f)^+$. The point $w_i = c_i^+$ for $i = 1, \ldots, m$ and $t = (\sum_{i:c_i>0} c_i + f)^+$ is clearly feasible in the dual. Thus we have found a point in X and a dual solution of the same value. ∎

Approach 5. Show that if $\pi x \leq \pi_0$ defines a facet of conv(X), then it must be identical to one of the inequalities $a^i x \leq b_i$ defining P.

Example 9.4 (cont) Consider the inequality $\sum_{i=1}^m \pi_i x_i + \pi_{m+1} y \leq \pi_0$. Let $S = \{i \in \{1, \ldots, m\} : \pi_i > 0\}$ and $T = \{i \in \{1, \ldots, m\} : \pi_i < 0\}$. Note that as the point $(\mathbf{0}, 0) \in X$, $\pi_0 \geq 0$, and as $(e^S, 1) \in X$, $\sum_{i \in S} \pi_i + \pi_{m+1} \leq \pi_0$, where e^S is the characteristic vector of S. Also a facet-defining inequality must have a tight point with $y = 1$. The point $(e^S, 1)$ maximizes the lhs, and so $\sum_{i \in S} \pi_i + \pi_{m+1} = \pi_0 \geq 0$.

Now consider the valid inequality obtained as a nonnegative combination of valid inequalities:

$x_i - y \leq 0$ with weight π_i for $i \in S$
$-x_i \leq 0$ with weight $-\pi_i$ for $i \in T$
$y \leq 1$ with weight $\sum_{i \in S} \pi_i + \pi_{m+1}$.

The resulting inequality is $\sum_{i=1}^m \pi_i x_i + \pi_{m+1} y \leq \sum_{i \in S} \pi_i + \pi_{m+1}$. This dominates or equals the original inequality as $\sum_{i \in S} \pi_i + \pi_{m+1} \leq \pi_0$. So the only inequalities that are not nonnegative combinations of other inequalities are those describing P. ∎

Approach 6. Show that for any $c \in R^n, c \neq 0$, the set of optimal solutions $M(c)$ to the problem $\max\{cx : x \in X\}$ lies in $\{x : a^i x = b_i\}$ for some $i = 1, \ldots, m$, where $a^i x \leq b_i$ for $i = 1, \cdots, m$ are the inequalities defining P.

Example 9.4 (cont) Consider an arbitrary objective $(c, f) \in R^m \times R^1$.

If $f > 0$, $y = 1$ in every optimal solution and so $M(c, f) \in \{(x, y) : y = 1\}$.
If $c_i < 0$, then $x_i = 0$ in every optimal solution.
If $c_i > 0$ and $f \leq 0$, then $x_i = y$ in every optimal solution.
If $c_i = 0$ for all i and $f < 0$, then $x_i = 0$ in any optimal solution.

All cases have been covered, and so $P = \text{conv}(X)$. ∎

Approach 7. Verify that $b \in Z^n$, and show that for all $c \in Z^n$, the optimal value of the dual w^{LP} is integer valued. This is to show that the inequalities $Ax \leq b$ form a TDI system, see Theorem 3.14.

Example 9.4 (cont) We have shown using Approach 4 that $w^{LP} = (\sum_{i:c_i>0} c_i + f)^+$. This is integer valued when c and f are integral. ∎

Approach 8. (Projection from an Extended Formulation). Suppose $Q \subseteq R^n \times R^p$ is a polyhedron with $P = proj_x(Q)$ as defined in Section 1.7. Show that for all $c \in R^n$, the linear program $\max\{cx : (x,w) \in Q\}$ has an optimal solution with $x \in X$.

Example 9.5 (Uncapacitated Lot-Sizing). It can be shown that solving the extended formulation presented in Section 1.6 as a linear program gives a solution with the set-up variables $y_1, \ldots, y_n$ integral, and thus provides an optimal solution to ULS. So its projection to the (x, y, s) space describes the convex hull of solutions to ULS. ∎

9.3 0–1 KNAPSACK INEQUALITIES

Consider the set $X = \{x \in B^n : \sum_{j=1}^n a_j x_j \leq b\}$. Complementing variables if necessary by setting $\bar{x}_j = 1 - x_j$, we assume throughout this section that the coefficients $\{a_j\}_{j=1}^n$ are positive. Also we assume $b > 0$. Let $N = \{1, \ldots, n\}$.

9.3.1 Cover Inequalities

Definition 9.6 A set $C \subseteq N$ is a *cover* if $\sum_{j \in C} a_j > b$. A cover is *minimal* if $C \setminus \{j\}$ is not a cover for any $j \in C$.

Note that C is a cover if and only if its associated incidence vector x^C is infeasible for S.

Proposition 9.3 *If $C \subseteq N$ is a cover, the cover inequality*

$$\sum_{j \in C} x_j \leq |C| - 1$$

is valid for X.

ie, if C is a cover then you can't take ALL the objects in C... you can potentially take at most $|C|-1$ of them.

Proof. We show that if x^R does not satisfy the inequality, then $x^R \notin X$. If $\sum_{j \in C} x_j^R > |C| - 1$, then $|R \cap C| = |C|$ and thus $R \supseteq C$. Then $\sum_{j=1}^n a_j x_j^R = \sum_{j \in R} a_j \geq \sum_{j \in C} a_j > b$ and so $x^R \notin X$. ∎

148 STRONG VALID INEQUALITIES

Example 9.6 Consider the knapsack set

$$X = \{x \in B^7 : 11x_1 + 6x_2 + 6x_3 + 5x_4 + 5x_5 + 4x_6 + x_7 \leq 19\}.$$

Some minimal cover inequalities for X are: $|C|-1$

$$
\begin{aligned}
x_1 + x_2 + x_3 &&&&&&& \leq 2 \\
x_1 + x_2 &&&&& + x_6 && \leq 2 \\
x_1 &&&& + x_5 & + x_6 && \leq 2 \\
& & x_3 + x_4 & + x_5 & + x_6 && \leq 3.
\end{aligned}
$$

9.3.2 Strengthening Cover Inequalities

Are the cover inequalities "strong"? Is it possible to strengthen the cover inequalities so that they provide better cuts?

First we observe that there is a simple way to strengthen the basic cover inequality.

> **Proposition 9.4** If C is a cover for X, the *extended cover inequality*
> $$\sum_{j \in E(C)} x_j \leq |C| - 1$$
> is valid for X, where $E(C) = C \cup \{j : a_j \geq a_i \text{ for all } i \in C\}$.

The proof of validity is almost identical to that of Proposition 9.3.

Example 9.6 (cont) The extended cover inequality for $C = \{3, 4, 5, 6\}$ is $x_1 + x_2 + x_3 + x_4 + x_5 + x_6 \leq 3$. So the cover inequality $x_3 + x_4 + x_5 + x_6 \leq 3$ is dominated by the valid inequality $x_1 + x_2 + x_3 + x_4 + x_5 + x_6 \leq 3$.

★ Observe, however, that this is in turn dominated by the inequality $2x_1 + x_2 + x_3 + x_4 + x_5 + x_6 \leq 3$. (Since if you choose x_1, you can only choose 1 more)

★ Can we define a procedure that allows us to find the last inequality, which is nonredundant (facet-defining) and thus as strong as possible?
(Check it, $F = \{x \in P : 2x_1 + x_2 + x_3 + x_4 + x_5 + x_6 = 3\}$ is a facet of Conv(X))

Example 9.6 (cont) Consider again the cover inequality for $C = \{3, 4, 5, 6\}$. Clearly when $x_1 = x_2 = x_7 = 0$, the inequality $x_3 + x_4 + x_5 + x_6 \leq 3$ is valid for $\{x \in B^4 : 6x_3 + 5x_4 + 5x_5 + 4x_6 \leq 19\}$.

Now keeping $x_2 = x_7 = 0$, we ask for what values of α_1, the inequality

$$\alpha_1 x_1 + x_3 + x_4 + x_5 + x_6 \leq 3$$

is valid for $\{x \in B^5 : 11x_1 + 6x_3 + 5x_4 + 5x_5 + 4x_6 \leq 19\}$?

When $x_1 = 0$, the inequality is known to be valid for all values of α_1.
When $x_1 = 1$, it is a valid inequality

if and only if $\alpha_1 + x_3 + x_4 + x_5 + x_6 \leq 3$ is valid for all $x \in B^4$ satisfying $6x_3 + 5x_4 + 5x_5 + 4x_6 \leq 19 - 11$, or equivalently
if and only if $\alpha_1 + \max\{x_3 + x_4 + x_5 + x_6 : 6x_3 + 5x_4 + 5x_5 + 4x_6 \leq 8, x \in B^4\} \leq 3$, or equivalently
if and only if $\alpha_1 \leq 3 - \zeta$, where $\zeta = \max\{x_3 + x_4 + x_5 + x_6 : 6x_3 + 5x_4 + 5x_5 + 4x_6 \leq 8, x \in B^4\}$.

Now $\zeta = 1$ at the point $x = (0, 0, 0, 1)$, and hence $\alpha_1 \leq 2$.

Thus the inequality is valid for all values of $\alpha_1 \leq 2$, and $\alpha_1 = 2$ gives the strongest inequality. ∎

In general the problem is to find best possible values for α_j for $j \in N \setminus C$ such that the inequality

$$\sum_{j \in C} x_j + \sum_{j \in N \setminus C} \alpha_j x_j \leq |C| - 1$$

is valid for X.

The procedure we now describe leads to such a set of values and in fact provides a facet-defining inequality for $\text{conv}(X)$ when C is a minimal cover and $a_j \leq b$ for all $j \in N$.

Procedure to Lift Cover Inequalities (Just Read ex. 9.6 below)

Let $j_1, \ldots, j_r$ be an ordering of $N \setminus C$. Set $t = 1$.

The valid inequality $\sum_{i=1}^{t-1} \alpha_{j_i} x_{j_i} + \sum_{j \in C} x_j \leq |C| - 1$ has been obtained so far. To calculate the largest value of α_{j_t} for which the inequality $\alpha_{j_t} x_{j_t} + \sum_{i=1}^{t-1} \alpha_{j_i} x_{j_i} + \sum_{j \in C} x_j \leq |C| - 1$ is valid, solve the knapsack problem:

$$\zeta_t = \max \sum_{i=1}^{t-1} \alpha_{j_i} x_{j_i} + \sum_{j \in C} x_j$$

$$\sum_{i=1}^{t-1} a_{j_i} x_{j_i} + \sum_{j \in C} a_j x_j \leq b - a_{j_t}$$

$$x \in \{0, 1\}^{|C|+t-1}.$$

Set $\alpha_{j_t} = |C| - 1 - \zeta_t$.
Stop if $t = r$.

It can be seen that ζ_t measures how much space is used up by the variables $\{j_1, \ldots, j_{t-1}\} \cup C$ in the lifted inequality when $x_{j_t} = 1$.

Example 9.6 (cont) If we take $C = \{3, 4, 5, 6\}, j_1 = 1, j_2 = 2$ and $j_3 = 7$, we have already been through the steps to calculate $\alpha_1 = 2$. Continuing with

variable x_2, $\zeta_2 =$
$$\max\{2x_1+x_3+x_4+x_5+x_6 : 11x_1+6x_3+5x_4+5x_5+4x_6 \leq 19-6 = 13, x \in B^5\}$$
$= 2$, and thus $\alpha_{j_2} = \alpha_2 = 3 - 2 = 1$.

A similar calculation shows that $\zeta_3 = 3$ and thus $\alpha_{j_3} = \alpha_7 = 0$. Thus we finish with a facet-defining inequality $2x_1+1x_2+1x_3+1x_4+1x_5+1x_6+0x_7 \leq 3$. ∎

Returning to the question of the strength of the cover inequalities, it is not difficult to show that $\sum_{j \in C} x_j \leq |C| - 1$ is facet-defining for $\operatorname{conv}(X')$ where $X' = \{x \in B^{|C|} : \sum_{j \in C} a_j x_j \leq b\}$. This suggests that the inequality is strong. On the other hand, it is clearly strengthened by the lifting procedure that terminates with a facet-defining inequality for $\operatorname{conv}(X)$.

9.3.3 Separation for Cover Inequalities

Now let $\mathcal{F}$ be the family of cover inequalities for X, and let us examine the separation problem for this family. Explicitly we are given a nonintegral point x^* with $0 \leq x_j^* \leq 1$ for all $j \in N$, and we wish to know whether x^* satisfies all the cover inequalities. To formalize this problem, note that the cover inequality can be rewritten as

$$\sum_{j \in C} x_j \leq |C|-1 \quad \Rightarrow \quad \sum_{j \in C}(1-x_j) \geq 1.$$

Thus it suffices to answer the question:

Does there exist a set $C \subseteq N$ with $\sum_{j \in C} a_j > b$ for which $\sum_{j \in C}(1-x_j^*) < 1$?, or put slightly differently,

Is $\zeta = \min_{C \subseteq N}\{\sum_{j \in C}(1-x_j^*) : \sum_{j \in C} a_j > b\} < 1$?

As the set C is unknown, this can be formulated as a 0–1 integer program where the variable $z_j = 1$ if $j \in C$ and $z_j = 0$ otherwise. So the question can be restated yet again,

Is $\zeta = \min\{\sum_{j \in N}(1-x_j^*)z_j : \sum_{j \in N} a_j z_j > b, z \in B^n\} < 1$?

Theorem 9.5 (i) *If $\zeta \geq 1$, x^* satisfies all the cover inequalities.*
(ii) *If $\zeta < 1$ with optimal solution z^R, the cover inequality $\sum_{j \in R} x_j \leq |R|-1$ cuts off x^* by an amount $1-\zeta$.*

Example 9.7 Consider the 0–1 knapsack set

$$X = \{x \in B^6 : 45x_1 + 46x_2 + 79x_3 + 54x_4 + 53x_5 + 125x_6 \leq 178\}.$$

and the fractional point $x^* = (0, 0, \frac{3}{4}, \frac{1}{2}, 1, 0)$. ← $1-x^*$

The cover separation problem is:

$$\min 1z_1 + 1z_2 + \tfrac{1}{4}z_3 + \tfrac{1}{2}z_4 + 0z_5 + 1z_6$$
$$45z_1 + 46z_2 + 79z_3 + 54z_4 + 53z_5 + 125z_6 > 178$$
$$z \in B^6.$$

An optimal solution is $z^R = (0, 0, 1, 1, 1, 0)$ with $\zeta = \frac{3}{4}$. Thus the inequality

$$x_3 + x_4 + x_5 \leq 2$$

is violated by $1 - \zeta = \frac{1}{4}$. $|C|-1$ (since $C = \{3,4,5\}$) ∎

9.4 MIXED 0–1 INEQUALITIES

Here we consider the mixed 0–1 set:

$$X = \{(x,y) \in R_+^n \times B^n : \sum_{j \in N_1} x_j \leq b + \sum_{j \in N_2} x_j, \; x_j \leq a_j y_j \text{ for } j \in N_1 \cup N_2\}.$$

Note that when $N_2 = \phi$ and $x_j = a_j y_j$ for all $j \in N_1$, this reduces to the knapsack set studied in Section 9.3. The set X can be viewed as the feasible region of a simple fixed charge flow network (see Figure 9.3).

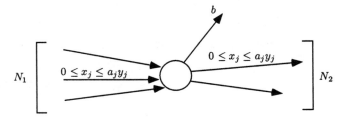

Fig. 9.3 Mixed 0–1 flow

9.4.1 Flow Cover Inequalities

Definition 9.7 *A set $C = C_1 \cup C_2$ with $C_1 \subseteq N_1, C_2 \subseteq N_2$ is a generalized cover for X if $\sum_{j \in C_1} a_j - \sum_{j \in C_2} a_j = b + \lambda$ with $\lambda > 0$. λ is called the (cover-) excess.*

Proposition 9.6 *The flow cover inequality*

$$\sum_{j \in C_1} x_j + \sum_{j \in C_1} (a_j - \lambda)^+ (1 - y_j) \leq b + \sum_{j \in C_2} a_j + \lambda \sum_{j \in L_2} y_j + \sum_{j \in N_2 \setminus (C_2 \cup L_2)} x_j$$

is valid for X, where $L_2 \subseteq N_2 \setminus C_2$.

Proof. Let $C_1^+ = \{j \in C_1 : a_j > \lambda\}$. Now given a point $(x,y) \in X$, we must show that (x,y) satisfies the inequality. Let $T = \{j \in N_1 \cup N_2 : y_j = 1\}$. There are two cases.

Case 1. $|C_1^+ \setminus T| + |L_2 \cap T| = 0$. Now

$$
\begin{aligned}
&\phantom{={}} \sum_{j \in C_1} x_j + \sum_{j \in C_1}(a_j - \lambda)^+(1 - y_j) \\
&= \sum_{j \in C_1 \cap T} x_j + \sum_{j \in C_1^+ \setminus T}(a_j - \lambda) \\
&= \sum_{j \in C_1 \cap T} x_j \quad (\text{as } |C_1^+ \setminus T| = 0) \\
&\leq \sum_{j \in N_1} x_j \quad (\text{as } x_j \geq 0) \\
&\leq b + \sum_{j \in N_2} x_j \quad (\text{by definition of } X) \\
&= b + \sum_{j \in C_2} x_j + \sum_{j \in L_2 \cap T} x_j + \sum_{j \in N \setminus (C_2 \cup L_2)} x_j \\
&\leq b + \sum_{j \in C_2} a_j + 0 + \sum_{j \in N \setminus (C_2 \cup L_2)} x_j \quad (\text{as } |L_2 \cap T| = 0) \\
&= b + \sum_{j \in C_2} a_j + \lambda \sum_{j \in L_2} y_j + \sum_{j \in N \setminus (C_2 \cup L_2)} x_j.
\end{aligned}
$$

Case 2. $|C_1^+ \setminus T| + |L_2 \cap T| \geq 1$.

$$
\begin{aligned}
&\phantom{={}} \sum_{j \in C_1} x_j + \sum_{j \in C_1}(a_j - \lambda)^+(1 - y_j) \\
&= \sum_{j \in C_1 \cap T} x_j + \sum_{j \in C_1^+ \setminus T}(a_j - \lambda) \\
&\leq \sum_{j \in C_1} a_j - |C_1^+ \setminus T|\lambda \quad (\text{as } x_j \leq a_j) \\
&\leq \sum_{j \in C_1} a_j - \lambda + \lambda|L_2 \cap T| \quad (\text{as } -|C_1^+ \setminus T| \leq -1 + |L_2 \cap T|) \\
&= b + \sum_{j \in C_2} a_j + \lambda \sum_{j \in L_2} y_j \\
&\leq b + \sum_{j \in C_2} a_j + \lambda \sum_{j \in L_2} y_j + \sum_{j \in N \setminus (C_2 \cup L_2)} x_j \quad (\text{as } x_j \geq 0).
\end{aligned}
$$

Example 9.8 Consider the set

$$X = \{(x,y) \in R_+^6 \times B^6 : x_1 + x_2 + x_3 \leq 4 + x_4 + x_5 + x_6,$$

$$x_1 \leq 3y_1, x_2 \leq 3y_2, x_3 \leq 6y_3, x_4 \leq 3y_4, x_5 \leq 5y_5, x_6 \leq 1y_6\}.$$

Taking $C_1 = \{1, 3\}$ and $C_2 = \{4\}$, $C = (C_1, C_2)$ is a generalized cover with $\lambda = 2$. Taking $L_2 = \{5\}$, the resulting flow cover inequality is

$$x_1 + x_3 + 1(1 - y_1) + 4(1 - y_3) \leq 7 + 2y_5 + x_6.$$

9.4.2 Separation for Flow Cover Inequalities

If one makes the assumption that $x_j = a_j y_j$ and $a_j \geq \lambda$ for all $j \in C_1$, and that $L_2 = N_2 \setminus C_2$, the flow cover inequality becomes:

$$\sum_{j \in C_1} a_j y_j + \sum_{j \in C_1} (a_j - \lambda)(1 - y_j) \leq b + \sum_{j \in C_2} a_j + \lambda \sum_{j \in N_2 \setminus C_2} y_j,$$

or after simplification,

$$\sum_{j \in C_1} a_j - \lambda \sum_{j \in C_1} (1 - y_j) \leq b + \sum_{j \in C_2} a_j + \lambda \sum_{j \in N_2 \setminus C_2} y_j,$$

or dividing by λ,

$$\sum_{j \in C_1} (1 - y_j) + \sum_{j \in N_2 \setminus C_2} y_j \geq 1,$$

or after rewriting,

$$\sum_{j \in C_1} (1 - y_j) - \sum_{j \in C_2} y_j \geq 1 - \sum_{j \in N_2} y_j.$$

This is nothing but the cover inequality for a 0–1 knapsack set with both positive and negative coefficients $\{x \in B^n : \sum_{j \in N_1} a_j y_j - \sum_{j \in N_2} a_j y_j \leq b\}$ where the cover (C_1, C_2) satisfies $\sum_{j \in C_1} a_j - \sum_{j \in C_2} a_j = b + \lambda$ with $\lambda > 0$.

This immediately suggests a separation heuristic for the flow cover inequalities. Letting z be the unknown incidence vector of $C = (C_1, C_2)$, consider the knapsack problem

$$\zeta = \min \sum_{j \in N_1} (1 - y_j^*) z_j - \sum_{j \in N_2} y_j^* z_j$$
$$\sum_{j \in N_1} a_j z_j - \sum_{j \in N_2} a_j z_j > b$$
$$z \in B^n.$$

Let z^C, the incidence vector of C, be an optimal solution.

Flow Cover Separation Heuristic. Take the cover $C = (C_1, C_2)$ obtained by solving this knapsack problem, and test whether

$$\sum_{j \in C_1} x_j^* + \sum_{j \in C_1} (a_j - \lambda)^+ (1 - y_j^*) > b + \sum_{j \in C_2} a_j + \lambda \sum_{j \in L_2} y_j^* + \sum_{j \in N_2 \setminus (C_2 \cup L_2)} x_j^*$$

where $L_2 = \{j \in N_2 \setminus C_2 : \lambda y_j^* < x_j^*\}$. If so, a violated flow cover inequality has been found.

Example 9.8 (cont) Suppose that a solution (x^*, y^*) is given with $x^* = (3, 0, 4, 3, 0, 0)$ and $y^* = (1, 0, \frac{2}{3}, 1, 0, 0)$. Solving the knapsack problem

$$\zeta = \min 0z_1 + 1z_2 + \tfrac{1}{3}z_3 - 1z_4 - 0z_5 - 0z_6$$
$$3z_1 + 3z_2 + 6z_3 - 3z_4 - 5z_5 - z_6 > 4$$
$$z \in B^6,$$

one obtains as solution $C_1 = \{1, 3\}$, $C_2 = \{4\}$, and $\lambda = 2$. The resulting flow cover inequality

$$x_1 + x_3 + 1(1 - y_1) + 4(1 - y_3) \leq 7 + x_5 + x_6$$

is violated by $\frac{4}{3}$. ∎

As in the previous section, it is natural to ask if the flow cover inequality of Proposition 9.6 is facet-defining, and if not to attempt to strengthen the inequality by lifting. A partial answer is that the inequality is facet-defining when $x_j = y_j = 0$ for $j \in N_2$ and C_1 is a minimal cover. However calculating valid lifting coefficients is more complicated because of the presence of the $x_j \leq a_j y_j$ constraints, so it becomes a computational question whether such lifting is worthwhile in practice.

9.5 THE OPTIMAL SUBTOUR PROBLEM

Consider a variant of the traveling salesman problem in which the salesman makes a profit f_j if he visits city (client) $j \in N$, he pays travel costs c_e if he traverses edge $e \in E$, but he is not obliged to visit all the cities. His subtour must start and end at city 1, and include at least two other cities. This problem is also known as the *prize collecting traveling salesman problem*, and is a subproblem of certain vehicle routing problems. It is $\mathcal{NP}$-hard.

Introducing binary edge variables: $x_e = 1$ if edge e lies on the subtour and $x_e = 0$ otherwise for $e \in E$; and binary node variables: $y_j = 1$ if city j lies on the subtour and $y_j = 0$ otherwise for $j \in N$, one obtains the formulation

$$\max -\sum_{e \in E} c_e x_e + \sum_{j \in N} f_j y_j \tag{9.1}$$
$$\sum_{e \in \delta(i)} x_e = 2y_i \text{ for } i \in N \tag{9.2}$$
$$\sum_{e \in E(S)} x_e \leq \sum_{i \in S \setminus \{k\}} y_i \text{ for } k \in S, S \subseteq N \setminus \{1\} \tag{9.3}$$
$$y_1 = 1 \tag{9.4}$$
$$x \in B^{|E|}, y \in B^{|N|}. \tag{9.5}$$

Constraints (9.2) ensure that if a node is visited, two adjacent edges are used to enter and leave, and (9.3) are *generalized subtour elimination* constraints (GSEC), which ensure that there is no subtour that does not contain node

1. Note that if we allow $x_e \in \{0,1,2\}$ for $e \in \delta(1)$, we also allow subtours consisting of node 1 and just one other node.

The question addressed here is not to find new inequalities, but how to deal with the exponential number of generalized subtour elimination constraints.

9.5.1 Separation for Generalized Subtour Constraints

Let $N' = N \setminus \{1\}$, and $E' = E \setminus \{\delta(1)\}$. To formalize the separation problem, we represent the unknown set $S \subseteq N'$ by binary variables z with $z_i = 1$ if $i \in S$. Now (x^*, y^*) violates the (k, S) generalized subtour inequality if

$$\sum_{e \in E'(S)} x_e^* > \sum_{i \in S \setminus \{k\}} y_i^*,$$

if and only if $\zeta_k > 0$ where

$$\zeta_k = \max\{\sum_{e=(i,j) \in E': i<j} x_e^* z_i z_j - \sum_{i \in N' \setminus \{k\}} y_i^* z_i : z \in B^{|N'|}, z_k = 1\}.$$

Note that this is a quadratic 0–1 program without constraints.

We now show how this problem can be solved in polynomial time. First we convert it into a linear integer program, and then we observe that the linear programming relaxation is integral.

We introduce variables w_e with $w_e = 1$ if $z_i = z_j = 1$ for $e = (i,j) \in E'$ with $i < j$, and we obtain the integer program

$$\zeta_k = \max \sum_{e \in E'} x_e^* w_e - \sum_{i \in N' \setminus \{k\}} y_i^* z_i \quad (9.6)$$

(IP) $\quad w_e \leq z_i, w_e \leq z_j \text{ for } e = (i,j) \in E' \quad (9.7)$

$\quad w_e \geq z_i + z_j - 1 \text{ for } e = (i,j) \in E' \quad (9.8)$

$\quad w \in B^{|E'|}, z \in B^{|N'|}, z_k = 1. \quad (9.9)$

Proposition 9.7 *If $x_e^* \geq 0$ for $e \in E'$, the linear program consisting of (9.6),(9.7) and the bound constraints obtained by relaxing integrality in (9.9) always has an integer optimal solution solving IP.*

Proof. Consider the linear programming relaxation of IP. As $x_e^* \geq 0$ for all $e \in E'$, there exists an optimal solution with w_e as large as possible. Thus by (9.7), $w_e = \min(z_i, z_j)$. As $z_i \leq 1$ for all $i \in N'$, $\min(z_i, z_j) \geq z_i + z_j - 1$, and so (9.8) is automatically satisfied, and can be dropped. Now each constraint of the linear programming relaxation contains two nonzero coefficients of -1 and $+1$, and is thus totally unimodular by Proposition 3.2 ∎

The constraint matrix in the dual of the linear program in Proposition 9.7 is a node-arc incidence matrix, and so the problem can be solved as a max flow problem. As $k \in \{2, \ldots, n\}$, the separation problem for GSECs can be

156 STRONG VALID INEQUALITIES

solved exactly by solving $n-1$ max flow problems. In practice, very fast separation heuristics have also been developed in computational studies of the TSP, which can be used to find many violated GSECs.

Example 9.9 We consider an instance of the optimal subtour problem with

$$n = 7, \ f = (2, 4, 1, 3, 7, 1, 7) \text{ and } (c_e) = \begin{pmatrix} - & 4 & 3 & 3 & 5 & 2 & 5 \\ - & - & 5 & 3 & 3 & 4 & 7 \\ - & - & - & 4 & 6 & 0 & 4 \\ - & - & - & - & 4 & 4 & 6 \\ - & - & - & - & - & 5 & 8 \\ - & - & - & - & - & - & 3 \\ - & - & - & - & - & - & - \end{pmatrix}.$$

Iteration 0. We solve a linear programming relaxation of problem (9.1)–(9.5) consisting of the degree constraints (9.2), constraint $y_1 = 1$, the trivial GSECs $x_e \leq y_i, x_e \leq y_j$ for $e = (i,j) \in E$, and the bounds on the variables.

The resulting solution consists of two subtours $(1, 5, 2, 4)$ and $(3, 6, 7)$ with $z^{LP} = 4$.

Iteration 1. The constraint $x_{36} + x_{37} + x_{67} \leq y_3 + y_7$ is added cutting off the subtour $(3, 6, 7)$.

The solution of the new relaxation consists of two subtours $(1, 3, 6, 7)$ and $(2, 4, 5)$ with $z^{LP} = 4$.

Iteration 2. The constraint $x_{24} + x_{25} + x_{45} \leq y_4 + y_5$ is added cutting off the subtour $(2, 4, 5)$.

The new LP solution is shown in Figure 9.4 with $z^{LP} = 4$.

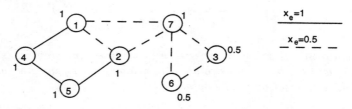

Fig. 9.4 Fractional optimal subtour solution

The separation problem for $k = 7$ is the linear program:

$$\zeta_7 = \max 1w_{45} + 1w_{25} + 0.5w_{27} + 0.5w_{67} + 0.5w_{37} + 0.5w_{36}$$
$$-1z_2 - 0.5z_3 - 1z_4 - 1z_5 - 0.5z_6$$

subject to $0 \leq w_e \leq z_i, z_j \leq 1$ for $e = (i,j) \in E'$, $z_7 = 1$.

An optimal solution is $z_3 = z_6 = z_7 = w_{36} = w_{37} = w_{67} = 1$ with violation $\zeta_7 = \frac{1}{2}$.

Iteration 3. The constraint $x_{36} + x_{37} + x_{67} \leq y_3 + y_6$ is added.

The new linear programming solution is the single subtour $(1,2,5,4,7,6)$, which is thus optimal with value of 2. ∎

9.6 BRANCH-AND-CUT

Successful solution of difficult IPs requires a combination of the many different ideas developed in this and earlier chapters. In Chapter 7, in discussing branch-and-bound, we mentioned the importance of efficient preprocessing, fast solution and reoptimization of linear programs, and good search strategies.

The need for good primal heuristics has been stressed, and ideas for using the linear programming solution to try to construct feasible solutions are discussed in Chapter 12. In this and the previous chapter we have also considered the use of separation algorithms to generate cutting planes which tighten up the formulations.

A *Branch-and-Cut Algorithm* is a branch-and-bound algorithm in which cutting planes are generated throughout the branch-and-bound tree. Though this may seem to be a minor difference, in practice there is a change of philosophy. Rather than reoptimizing fast at each node, the new philosophy is to do as much work as is necessary to get a tight dual bound at the node. Now the goal is not only to reduce the number of nodes in the tree significantly by using cuts and improved formulations, but also by trying anything else that may be useful such as preprocessing at each node, a primal heuristic at each node, and so forth.

In practice there is obviously a trade-off. If many cuts are added at each node, reoptimization may be much slower than before. In addition, keeping all the information in the tree is significantly more difficult. In branch-and-bound the problem to be solved at each node is obtained just by adding bounds. In branch-and-cut a *cut pool* is used in which all the cuts are stored. In addition to keeping the bounds and a good basis in the node list, it is also necessary to indicate which constraints are needed to reconstruct the formulation at the given node, so pointers to the appropriate constraints in the cut pool are kept.

A flowchart of the basic steps of a branch-and-cut algorithm is shown in Figure 9.5.

Example 9.10 Generalized Assignment Problem. We consider an instance of the problem

$$\max \sum_{i=1}^{m} \sum_{j=1}^{n} c_{ij} x_{ij}$$
$$\sum_{j=1}^{n} x_{ij} \leq 1 \text{ for } i = 1, \ldots, m$$
$$\sum_{i=1}^{m} a_{ij} x_{ij} \leq b_j \text{ for } j = 1, \ldots, n$$
$$x \in B^{mn}$$

STRONG VALID INEQUALITIES

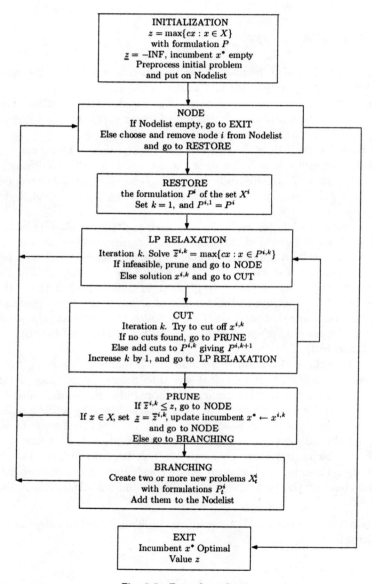

Fig. 9.5 Branch-and-cut

with $m = 10, n = 5, b = (91, 87, 109, 88, 64)^T$,

$$(c_{ij}) = \begin{pmatrix} 110 & 16 & 25 & 78 & 59 \\ 65 & 69 & 54 & 28 & 71 \\ 19 & 93 & 45 & 45 & 9 \\ 89 & 31 & 72 & 83 & 20 \\ 62 & 17 & 77 & 18 & 39 \\ 37 & 115 & 87 & 59 & 97 \\ 89 & 102 & 98 & 74 & 61 \\ 78 & 96 & 87 & 55 & 77 \\ 74 & 27 & 99 & 91 & 5 \\ 88 & 97 & 99 & 99 & 51 \end{pmatrix}, (a_{ij}) = \begin{pmatrix} 95 & 1 & 21 & 66 & 59 \\ 54 & 53 & 44 & 26 & 60 \\ 3 & 91 & 43 & 42 & 5 \\ 72 & 30 & 56 & 72 & 9 \\ 44 & 1 & 71 & 13 & 27 \\ 20 & 99 & 87 & 52 & 85 \\ 72 & 96 & 97 & 73 & 49 \\ 75 & 82 & 83 & 44 & 59 \\ 68 & 8 & 87 & 74 & 4 \\ 69 & 83 & 98 & 88 & 45 \end{pmatrix}.$$

We use a branch-and-cut system generating lifted cover inequalities. Three passes are made at the top node with up to 5 cuts added per pass. One pass is made at all other tree nodes with up to 5 cuts added per pass. The resulting branch-and-cut tree is shown in Figure 9.6.

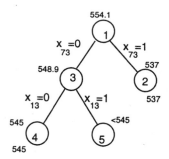

Fig. 9.6 Branch-and-cut tree

Initialization. The preprocessor eliminates the five variables $x_{11}, x_{32}, x_{62}, x_{72}, x_{65}$ with $a_{ij} > b_j$.

Node 1. The initial linear programming value is 595.6. After three passes, 14 cuts have been generated, and the improved linear programming value is 554.1, with corresponding linear programming solution: $x_{12} = 1, x_{23} = x_{25} = 0.5, x_{31} = 0.76, x_{33} = 0.24, x_{43} = x_{45} = 0.5, x_{52} = 0.74, x_{54} = 0.26, x_{61} = 1, x_{73} = x_{75} = 0.5, x_{82} = 1, x_{91} = 0.76, x_{94} = 0.24, x_{10,1} = 0.24, x_{10,4} = 0.76, x_{ij} = 0$ otherwise.

x_{73} is chosen as branching variable, creating node 2 with $x_{73} = 1$ and node 3 with $x_{73} = 0$.

Node 2. Three cuts are added. The resulting linear programming solution is integral with value 537. The incumbent value $\underline{z}$ and solution x^* are updated. The node is pruned by optimality.

160   STRONG VALID INEQUALITIES

Node 3. One cut is generated. The resulting linear programming solution with value 548.9 is nonintegral. x_{13} is chosen as branching variable, creating node 4 with $x_{13} = 0$ and node 5 with $x_{13} = 1$.

Node 4. The linear programming solution is integral with value 545. The incumbent value and solution are updated. The node is pruned by optimality.

Node 5. The linear programming solution value does not exceed the incumbent value of 545. The node is pruned by bound.

Termination. The node list is empty. $z = 545$ with optimal solution $x_{12} = x_{23} = x_{31} = x_{43} = x_{52} = x_{61} = x_{75} = x_{82} = x_{91} = x_{10,4} = 1$ and $x_{ij} = 0$ otherwise.

If the number of cut passes is unlimited at the top node, 18 cuts are generated, the final linear programming value is 546, and the branch-and-cut tree has just three nodes. Using just branch-and-bound, 4206 nodes are required. ∎

9.7 NOTES

The use of cutting planes as a practical tool for solving general integer programs was almost completely abandoned in the 1960s and 1970s. However, the interest in looking for strong valid inequalities for $\mathcal{NP}$-hard problems in the early 1970s can perhaps be traced to two sources:
(i) The beautiful polyhedral results of Edmonds and Fulkerson in the late 1960s in deriving the convex hulls of the sets of incidence vectors for matching, branching, and matroid problems, as well as the results of Gomory, and Gomory and Johnson, in developing characterizations of the valid inequalities for group and integer programming problems, and
(ii) The idea that because of $\mathcal{NP}$-completeness theory, there is no hope of an efficient algorithm for integer programming, and therefore that it is important to study specific problem structure if progress on $\mathcal{NP}$-hard problems is to be made.
The first polyhedral studies in the early seventies, other than those of the knapsack polytope, were of the stable set polytope [Pad73], [NemTro74] and the *TSP* polytope [GroPad75].

9.2 Books on polyhedra include [Gru67], [StoWit70], and [Zie95]. Surveys specially written with integer programming and combinatorial optimization in view are [Pul83] and Chapter I.1.4 in [NemWol88].

9.3 Cover inequalities for 0–1 knapsack polytopes were developed simultaneously in [Balas75b], [HamJohPel75], and [Wol75a]. The concept of lifting, already present in [Gom69], was extended in [Pad73], [Wol75b]. A major step forward came with the use of these inequalities in a cutting plane/branch-and-bound algorithm for 0–1 *IPs* [CroJohPad83]. Twenty years later separation routines for these inequalities are finally making their way into the commercial codes such as CPLEX, XOSL, and XPRESS.

Recent developments include the derivation of larger classes of inequalities [Wei97] for 0–1 knapsacks, and the generalization of cover inequalities to 0–1 knapsacks with GUB constraints [Wol90], [Guetal95], and to integer knapsacks [Cerietal95].

9.4 Flow cover inequalities appeared first in [PadVanRWol85] and were implemented computationally in [VRoWol87]. Other related inequalities for the model appear in [Sta92]. Recently stronger liftings of the flow cover inequalities and extensive computational tests are presented in [Guetal96]. A new mixed integer model consisting of a 0–1 knapsack constraint plus a single continuous variable is studied in [MarWol97].

9.5 The optimal subtour problem is studied in [Balas89]; see also [Bau97]. The GSEC inequalities are proposed by [Goe94] in studying the Steiner tree problem. The reduction of the quadratic 0–1 problem to the dual of a network flow problem is folklore. The reduction to a series of max flow problems is in [Rhy70]. The problem of finding violated subtour elimination constraints quickly is faced by all the researchers developing computational studies of the *TSP* problem. [PadRin91] present very effective heuristic separation routines. A recent survey on *TSP* is [JueReiRin95].

9.6 Two recent surveys on branch-and-cut are [JueReiThi95] and [LucBea96]. See also the annotated bibliography [CapFis97] which contains references to applications in general mixed integer programming, routing, scheduling, graph partitioning, set packing and covering, network design, and location problems among others.

Three research codes especially built to allow the easy development of branch-and-cut applications are MINTO [NemSavSig94], ABACUS [Thi95] and *bc − opt* [Cordetal97].

9.8 EXERCISES

1. Consider the set $X = \{x \in Z_+^2 : x_1 - x_2 \geq -1, 2x_1 + 6x_2 \leq 15, x_1 - x_2 \leq 3, 2x_1 + 4x_2 \leq 7\}$. List and represent graphically the set of feasible points. Use this to find a minimal description of conv(X).

2. Let $X = \{x \in B^n : \sum_{j \in N} a_j x_j \leq b\}$ with $a_j > 0$ for $j \in N$ and $b > 0$. Show that a valid inequality $\sum_{j \in N} \pi_j x_j \leq \pi_0$ with $\pi_0 > 0$ and $\pi_j < 0$ for $j \in T \neq \emptyset$ is dominated by the valid inequality $\sum_{j \in N} \max[\pi_j, 0] x_j \leq \pi_0$.

STRONG VALID INEQUALITIES

3. In each of the examples below a set X and a point x^* are given. Find a valid inequality for X cutting off x^*.

(i)
$$X = \{x \in B^5 : 9x_1 + 8x_2 + 6x_3 + 6x_4 + 5x_5 \leq 14\}$$
$$x = (0, \frac{5}{8}, \frac{3}{4}, \frac{3}{4}, 0),$$

(ii)
$$X = \{x \in B^5 : 9x_1 + 8x_2 + 6x_3 + 6x_4 + 5x_5 \leq 14\}$$
$$x = (\frac{1}{2}, \frac{1}{8}, \frac{3}{4}, \frac{3}{4}, 0),$$

(iii)
$$X = \{x \in B^5 : 7x_1 + 6x_2 + 6x_3 + 4x_4 + 3x_5 \leq 14\}$$
$$x = (\frac{1}{7}, 1, \frac{1}{2}, \frac{1}{4}, 1),$$

(iv)
$$X = \{x \in B^5 : 12x_1 - 9x_2 + 8x_3 + 6x_4 - 3x_5 \leq 2\}$$
$$x = (0, 0, \frac{1}{2}, \frac{1}{6}, 1).$$

4. Consider the knapsack set
$$X = \{x \in B^6 : 12x_1 + 9x_2 + 7x_3 + 5x_4 + 5x_5 + 3x_6 \leq 14\}.$$

Set $x_1 = x_2 = x_4 = 0$, and consider the cover inequality $x_3 + x_5 + x_6 \leq 2$ that is valid for $X' = X \cap \{x : x_1 = x_2 = x_4 = 0\}$.

(i) Lift the inequality to obtain a strong valid inequality $\alpha_1 x_1 + \alpha_2 x_2 + \alpha_4 x_4 + x_3 + x_5 + x_6 \leq 2$ for X.

(ii*) Show that $x_3 + x_5 + x_6 \leq 2$ defines a facet of X'.
(iii*) Use (ii) and the calculations in (i) to show that the inequality obtained in (i) defines a facet of $\text{conv}(X)$.

5. Consider the set $X = \{(x, y) \in \mathbb{R}_+^4 \times B^4 : x_1 + x_2 + x_3 + x_4 \geq 36, x_1 \leq 20y_1, x_2 \leq 10y_2, x_3 \leq 10y_3, x_4 \leq 8y_4\}$

(i) Derive a valid inequality that is a 0–1 knapsack constraint.
(ii) Use this to cut off the fractional point $x^* = (20, 10, 0, 6), y^* = (1, 1, 0, \frac{3}{4})$ with an inequality involving only y variables.

6. In each of the examples below a set X and a point (x^*, y^*) are given. Find a valid inequality for X cutting off (x^*, y^*).

(i) $X = \{(x,y) \in R_+^3 \times B^3 : x_1 + x_2 + x_3 \leq 7, x_1 \leq 3y_1, x_2 \leq 5y_2, x_3 \leq 6y_3\}$
$(x^*, y^*) = (2, 5, 0, \frac{2}{3}, 1, 0)$.

(ii) $X = \{(x,y) \in R_+^3 \times B^3 : 7 \leq x_1 + x_2 + x_3, x_1 \leq 3y_1, x_2 \leq 5y_2, x_3 \leq 6y_3\}$
$(x^*, y^*) = (2, 5, 0, \frac{2}{3}, 1, 0)$.

(iii) $X = \{(x,y) \in R_+^6 \times B^6 : x_1 + x_2 + x_3 \leq 4 + x_4 + x_5 + x_6, x_1 \leq 3y_1, x_2 \leq 3y_2, x_3 \leq 6y_3, x_4 \leq 3y_4, x_5 \leq 5y_5, x_6 \leq 1y_6\}$.

$(x^*, y^*) = (3, 3, 0, 0, 2, 0; 1, 1, 0, 0, \frac{2}{5}, 0)$.

7. Consider the set $X = \{(x, y_2) \in R_+^3 \times B^1 : x_1 + x_2 \leq b + x_3, x_2 \leq My_2\}$. Derive a flow cover inequality with $C = (\{2\}, \emptyset)$. For the constraints of ULS, $s_{t-1} + x_t = d_t + s_t, x_t \leq My_t, x_t, s_{t-1}, s_t \geq 0, y_t \in B^1$, what is the resulting inequality?

8. Consider the set $X = \{(x_1, y_2, y_3, y_4) \in R_+^1 \times B^3 : x_1 \leq 1000y_2 + 1500y_3 + 2200y_4, x_1 \leq 2000\}$.
Show that it can be rewritten as a mixed 0–1 set

$$X' = \{(x, y) \in R_+^4 \times B^4 : x_1 \leq x_2 + x_3 + x_4,$$

$$x_1 \leq 2000y_1, x_2 \leq 1000y_2, x_3 \leq 1500y_3, x_4 \leq 2200y_4\}$$

with the additional constraints

$$y_1 = 1, x_2 \geq 1000y_2, x_3 \geq 1500y_3, x_4 \geq 2200y_4.$$

Use this to find a valid inequality for X cutting off

$$(x_1, y_2, y_3, y_4) = (2000, 0, 1, \frac{5}{22}).$$

9. X is the set of incidence vectors of $STSP$ tours. Find a valid inequality for conv(X) cutting off the point x^* given in Figure 9.7.

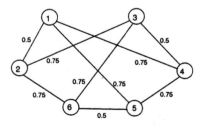

Fig. 9.7 Fractional solution of $STSP$

10. Solve the optimal subtour problem with $n = 6, f = (0, 19, 13, 1, 8, 12)$ and

$$(c_e) = \begin{pmatrix} - & 2 & 8 & 16 & 3 & 4 \\ & - & 6 & 4 & 3 & 8 \\ & & - & 10 & 12 & 9 \\ & & & - & 5 & 7 \\ & & & & - & 13 \\ & & & & & - \end{pmatrix}.$$

11. Solve the quadratic 0–1 problem

$$\max_{x \neq 0}\{x \in B^5 : x_1x_2 + 2x_1x_3 + 3x_1x_4 + 8x_1x_5 + 1x_2x_4 + 3x_2x_5 + 6x_3x_4$$

$$+ 5x_4x_5 - x_1 - 4x_2 - 10x_3 - 2x_4 - 2x_5\}.$$

12. Consider the optimal subtree of a tree problem (Section 5.3) with an additional budget constraint, with feasible region: $X = \{x \in B^n : x_j \leq x_{p(j)}$ for $j \neq 1, \sum_{j \in N} a_j x_j \leq b\}$ where $p(j)$ is the predecessor of j in the tree rooted at node 1. $C \subseteq N$ is a *tree cover* if the nodes of C form a subtree rooted at node 1 and $\sum_{j \in C} a_j > b$.

Show that if C is a tree cover, the inequality $\sum_{j \in C}(x_{p(j)} - x_j) \geq 1$ is valid for X where $x_{p(1)} = 1$.

13. A possible formulation of the capacitated facility location problem is:

$$\sum_{j \in N} x_{ij} = a_i \text{ for } i \in M$$
$$\sum_{i \in M} x_{ij} \leq b_j y_j \text{ for } j \in N$$
$$x \geq R_+^{mn}, y \in B^n.$$

Given an instance with $m = 5$ and $n = 3$, use the substitution $v_j = \sum_{i \in M} x_{ij}$,

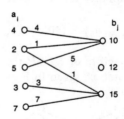

Fig. 9.8 Fractional solution of CFL

and $\sum_{j \in N} v_j = \sum_{i \in N} a_i$ to obtain an inequality cutting off the fractional solution consisting of the flow x^* shown in Figure 9.8 and $y^* = (1, \frac{4}{15}, \frac{11}{15})$.

14. Given a graph $G = (V, E)$ with $n = |V|$, consider the set of incidence vectors of the stable sets $X = \{x \in B^n : x_i + x_j \leq 1 \text{ for } e = (i,j) \in E\}$. Show that the clique inequality $\sum_{j \in C} x_j \leq 1$ is valid and defines a facet of conv(X), where a *clique* $C \subseteq V$ is a maximal set of nodes with the property

that for all pairs $i, j \in C$, the edge $e = (i,j)$ is in E.

15.* Show that inequalities at least as strong as the cover and extended cover inequalities can be obtained by one application of the Chvátal-Gomory procedure of Section 8.3.

16. Consider an instance of the generalized assignment problem

$$\max \sum_{i=1}^{m} \sum_{j=1}^{n} c_{ij} x_{ij}$$
$$\sum_{j=1}^{n} x_{ij} = 1 \text{ for } i = 1, \ldots, m$$
$$\sum_{i=1}^{m} a_{ij} x_{ij} \leq b_j \text{ for } j = 1, \ldots, n$$
$$x \in B_+^{mn}$$

with $m = 10, n = 5, b = (80, 63, 75, 98, 59)$ and

$$(a_{ij}) = \begin{pmatrix} 3 & 24 & 53 & 27 & 17 \\ 15 & 23 & 43 & 74 & 23 \\ 54 & 43 & 27 & 21 & 36 \\ 92 & 83 & 45 & 35 & 23 \\ 19 & 10 & 33 & 43 & 12 \\ 91 & 55 & 32 & 26 & 23 \\ 15 & 25 & 35 & 37 & 28 \\ 47 & 43 & 35 & 32 & 37 \\ 34 & 23 & 52 & 46 & 43 \\ 35 & 23 & 34 & 25 & 40 \end{pmatrix}, (c_{ij}) = \begin{pmatrix} 15 & 44 & 76 & 43 & 34 \\ 19 & 23 & 45 & 46 & 34 \\ 10 & 6 & 3 & 23 & 15 \\ 60 & 45 & 34 & 36 & 23 \\ 21 & 12 & 34 & 44 & 10 \\ 67 & 65 & 34 & 20 & 37 \\ 23 & 34 & 44 & 47 & 32 \\ 23 & 25 & 32 & 15 & 27 \\ 15 & 13 & 23 & 24 & 34 \\ 10 & 15 & 23 & 12 & 13 \end{pmatrix}.$$

Solve with a mixed integer programming system. Add valid inequalities so as to reduce the number of nodes.

10
Lagrangian Duality

10.1 LAGRANGIAN RELAXATION

Consider the integer program:

$$(IP) \quad \begin{aligned} z = \max\, & cx \\ & Ax \le b \quad \leftarrow \text{easy} \\ & Dx \le d \quad \leftarrow \text{complicated} \\ & x \in Z_+^n. \end{aligned}$$

Suppose that the constraints $Ax \le b$ are "nice" in the sense that an integer program with just these constraints (e.g., network constraints) is easy. Thus if one drops the "complicating constraints" $Dx \le d$, the resulting relaxation is easier to solve than the original problem IP. Many problems have such a structure, for example, the traveling salesman problem if one drops the connectivity constraints, the uncapacitated facility location problem if one drops the client demand constraints, and so on. However, the resulting bound obtained from the relaxation may be weak, because some important constraints are totally ignored. One way to tackle this difficulty is by *Lagrangian relaxation*, briefly introduced in Chapter 2.

We consider the problem IP in a slightly more general form:

$$\begin{aligned} z = \max\, & cx \\ & Dx \le d \quad \leftarrow \text{complicated} \\ & x \in X \quad \leftarrow \text{optimizing over this set is easy} \end{aligned}$$

where $Dx \le d$ are m complicating constraints.

LAGRANGIAN DUALITY

Note: d − Dx so that it stays a relaxation

For any value of $u = (u_1, \ldots, u_m) \geq 0$, we define the problem:

$$(IP(u)) \qquad z(u) = \max cx + u(d - Dx)$$
$$x \in X.$$

Proposition 10.1 *Problem $IP(u)$ is a relaxation of problem IP for all $u \geq 0$.*

Proof. Remember that $IP(u)$ is a relaxation of IP if:
(i) The feasible region is at least as large. This holds because $\{x : Dx \leq d, x \in X\} \subseteq X$.
(ii) The objective value is at least as great in $IP(u)$ as in IP for all feasible solutions in IP. As $u \geq 0$ and $Dx \leq d$ for all $x \in X$, $cx + u(d - Dx) \geq cx$ for all $x \in X$. ∎

We see that in $IP(u)$ the complicating constraints are handled by adding them to the objective function with a penalty term $u(d - Dx)$, or in other words, u is the *price* or *dual variable* or *Lagrange multiplier* associated with the constraints $Dx \leq d$.

Problem $IP(u)$ is called a *Lagrangian relaxation (subproblem)* of IP with parameter u. As $IP(u)$ is a relaxation of IP, $z(u) \geq z$ and we obtain an upper bound on the optimal value of IP. To find the best (smallest) upper bound over the infinity of possible values for u, we need to solve the *Lagrangian Dual Problem*:

$$(LD) \qquad w_{LD} = \min\{z(u) : u \geq 0\}.$$

• *Find the tightest upper bound.*

Observation 10.1 When the m constraints that are dualized are equality constraints of the form $Dx = d$, the corresponding Lagrange multipliers $u \in R^m$ are unrestricted in sign, and the Lagrangian dual becomes

$$w_{LD} = \min_u z(u).$$

Solving the Lagrangian relaxation $IP(u)$ may sometimes lead to an optimal solution of the original problem IP. *(Solving the relaxation $IP(u)$, not the Lagrangian Dual LD)*

KKT

Proposition 10.2 *If $u \geq 0$,*
(i) *$x(u)$ is an optimal solution of $IP(u)$, and*
(ii) *$Dx(u) \leq d$, and*
(iii) *$(Dx(u))_i = d_i$ whenever $u_i > 0$ (complementarity),*
then $x(u)$ is optimal in IP.

Proof. *Suppose these above conditions hold for some $u \geq 0$ and $x(u)$.* By (i) $w_{LD} \leq z(u) = cx(u) + u(d - Dx(u))$. By (iii) $cx(u) + u(d - Dx(u)) = cx(u)$. By (ii) $x(u)$ is feasible in IP and so $cx(u) \leq z$. Thus $w_{LD} \leq cx(u) + u(d - Dx(u)) = cx(u) \leq z$. But as $w_{LD} \geq z$, equality holds throughout, and $x(u)$ is optimal in IP. ∎

→ *That is, in the special case that KKT holds at $(x(u), u)$, we have found the optimum and needn't solve LD.*

Note also that if the constraints dualized are equality constraints, condition (iii) is automatically satisfied, and an optimal solution to $IP(u)$ is optimal for IP if it is feasible in IP.

Consider now the application of this approach to the uncapacitated facility location problem, starting with the strong formulation:

UFL:

(IP)
$$z = \max \sum_{i \in M} \sum_{j \in N} c_{ij} x_{ij} - \sum_{j \in N} f_j y_j$$
$$\sum_{j \in N} x_{ij} = 1 \text{ for } i \in M \quad \text{(set cover constraint, hard)}$$
$$x_{ij} - y_j \leq 0 \text{ for } i \in M, j \in N$$
$$x \in R^{|M| \times |N|}, y \in B^{|N|}.$$

Dualizing the demand constraints gives:

Fix j and we get a subproblem for each j

(IP(u))
$$z(u) = \max \sum_{i \in M} \sum_{j \in N} (c_{ij} - u_i) x_{ij} - \sum_{j \in N} f_j y_j + \sum_{i \in M} u_i$$
$$x_{ij} - y_j \leq 0 \text{ for } i \in M, j \in N$$
$$x \in R^{|M| \times |N|}, y \in B^{|N|}.$$

This in turn breaks up into a subproblem for each location. Thus $z(u) = \sum_{j \in N} z_j(u) + \sum_{i \in M} u_i$ where *(see p. 186)*

($IP_j(u)$)
$$z_j(u) = \max \sum_{i \in M} (c_{ij} - u_i) x_{ij} - f_j y_j$$
$$x_{ij} - y_j \leq 0 \text{ for } i \in M,$$
$$x_{ij} \geq 0 \text{ for } i \in M, y_j \in B^1.$$

Problem $IP_j(u)$ is easily solved by inspection. If $y_j = 0$, then $x_{ij} = 0$ for all i, and the objective value is 0. If $y_j = 1$, all clients that are profitable are served, namely those with $c_{ij} - u_i > 0$. The objective value is then $\sum_{i \in M} \max[c_{ij} - u_i, 0] - f_j$. So $z_j(u) = \max\{0, \sum_{i \in M} \max[c_{ij} - u_i, 0] - f_j\}$.

Example 10.1 Uncapacitated Facility Location. Consider an instance with $m = 6$ clients and $n = 5$ potential locations, fixed location costs $f = (2, 4, 5, 3, 3)$ and the client-location profit matrix

$$(c_{ij}) = \begin{pmatrix} 6 & 2 & 1 & 3 & 5 \\ 4 & 10 & 2 & 6 & 1 \\ 3 & 2 & 4 & 1 & 3 \\ 2 & 0 & 4 & 1 & 4 \\ 1 & 8 & 6 & 2 & 5 \\ 3 & 2 & 4 & 8 & 1 \end{pmatrix}.$$

Taking $u = (5, 6, 3, 2, 5, 4)$, we obtain the revised profit matrix

170   LAGRANGIAN DUALITY

$$(c_{ij} - u_i) = \begin{pmatrix} 1 & -3 & -4 & -2 & 0 \\ -2 & 4 & -4 & 0 & -5 \\ 0 & -1 & 1 & -2 & 0 \\ 0 & -2 & 2 & -1 & 2 \\ -4 & 3 & 1 & -3 & 0 \\ -1 & -2 & 0 & 4 & -3 \end{pmatrix}$$

with $\sum_{i \in M} u_i = 25$. The resulting Lagrangian problem $IP(u)$ is then solved by inspection as described above. For instance, for $j = 2$, we obtain value 0 if $y_2 = 0$. On the other hand, if $y_2 = 1$ at a cost of 4, we can set $x_{22} = 1$ as $c_{22} - u_2 = 4 > 0$ and $x_{52} = 1$ as $c_{52} - u_5 = 3$, and thus the net profit with $y_2 = 1$ is $7 - 4 = 3$. Hence it is optimal to set $y_2 = 1$ giving $z_2(u) = 3$.

Carrying out a similar calculation for each depot, an optimal solution of $IP(u)$ is to set $y_1 = y_3 = y_5 = 0, y_2 = x_{22} = x_{52} = 1, y_4 = x_{64} = 1$ giving $z(u) = 3 + 1 + \sum_{i \in M} u_i = 29$.   ∎

Now consider the application of Lagrangian relaxation to the symmetric traveling salesman problem. Remember that a *1-tree* is a tree on the graph induced by nodes $\{2, \ldots, n\}$ plus two edges incident to node 1; see Section 2.3. So a tour is any 1-tree having two edges incident to each node.

Alternatively, we have seen that the problem of finding a minimum cost tour can be formulated as an integer program:

$$z = \min \sum_{e \in E} c_e x_e \tag{10.1}$$
$$\sum_{e \in \delta(i)} x_e = 2 \text{ for all } i \in V \tag{10.2}$$
$$\sum_{e \in E(S)} x_e \leq |S| - 1 \text{ for all } 2 \leq |S| \leq |V| - 1 \tag{10.3}$$
$$x \in B^{|E|}. \tag{10.4}$$

Observation 10.2 Half the subtour constraints (10.3) are redundant. For any feasible solution of the linear programming relaxation of this formulation, $|S| - \sum_{e \in E(S)} x_e = \frac{1}{2} \sum_{i \in S} \sum_{e \in \delta(i)} x_e - \sum_{e \in E(S)} x_e = \frac{1}{2} \sum_{e \in \delta(S, \bar{S})} x_e$ where $\delta(S, \bar{S})$ is the set of edges with one endpoint in S and the other in $\bar{S} = V \setminus S$. Therefore, as $\delta(S, \bar{S}) = \delta(\bar{S}, S)$, $|S| - \sum_{e \in E(S)} x_e = |\bar{S}| - \sum_{e \in E(\bar{S})} x_e$ and so $\sum_{e \in E(S)} x_e \leq |S| - 1$ if and only if $\sum_{e \in E(\bar{S})} x_e \leq |\bar{S}| - 1$. Note also that summing the constraints (10.2) and dividing by 2, we obtain $\sum_{e \in E} x_e = n$.

Therefore, we can drop all subtour elimination constraints with $1 \in S$. Now to obtain a Lagrangian relaxation, we dualize all the degree constraints on the nodes, but also leave the degree constraint on node 1, and the constraint that the total number of edges is n, giving:

$$z(u) = \min \sum_{e \in E} (c_e - u_i - u_j) x_e + 2 \sum_{i \in V} u_i \tag{10.5}$$
$$\sum_{e \in \delta(1)} x_e = 2 \tag{10.6}$$

$$(IP(u)) \quad \sum_{e \in E(S)} x_e \leq |S| - 1 \text{ for } 2 \leq |S| \leq |V| - 1, \ 1 \notin S \quad (10.7)$$
$$\sum_{e \in E} x_e = n \quad (10.8)$$
$$x \in B^{|E|}. \quad (10.9)$$

The feasible solutions of $IP(u)$ are precisely the 1-trees. Finding a minimum weight 1-tree is easy, so this is potentially an interesting relaxation. Note also that as $STSP$ is a minimization problem, its Lagrangian dual is a maximization problem. As the dualized constraints are equality constraints, the dual variables are unrestricted in sign. Thus here $w_{LD} = \max_u z(u)$.

Example 10.2 Consider an instance of $STSP$ with edge cost matrix

$$(c_e) = \begin{pmatrix} - & 30 & 26 & 50 & 40 \\ - & - & 24 & 40 & 50 \\ - & - & - & 24 & 26 \\ - & - & - & - & 30 \\ - & - & - & - & - \end{pmatrix}.$$

Taking dual variables $u = (0, 0, -15, 0, 0, 0)$ and writing $\bar{c}_e = c_e - u_i - u_j$, we obtain the revised cost matrix

$$(\bar{c}_e) = \begin{pmatrix} - & 30 & 41 & 50 & 40 \\ - & - & 39 & 40 & 50 \\ - & - & - & 39 & 41 \\ - & - & - & - & 30 \\ - & - & - & - & - \end{pmatrix}.$$

The optimal 1-tree is found by taking the two cheapest edges out of node 1, that is, edges (1, 2) and (1, 5), plus the edges of an optimal tree on nodes $\{2, 3, 4, 5\}$. Choosing greedily, the edges (4, 5), (2, 3), and (3, 4) are chosen in that order. $z(u)$ is the sum of the length of this 1-tree and the value of $2 \sum u_i$, namely $178 - 30 = 148$. The resulting 1-tree is shown in Figure 10.1.

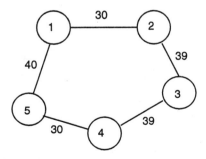

Fig. 10.1 Optimal tour for $STSP$

We have that $z(u) = 148 \leq z$. However, as the 1-tree is a tour it is readily

172 LAGRANGIAN DUALITY

checked that its length is also 148. Thus this tour is optimal. This also follows directly from Proposition 10.2. ∎

The questions that we would now like to answer are:

(i) How good is the upper bound obtained by solving the Lagrangian dual?

and

(ii) How can one solve the Lagrangian dual?

10.2 THE STRENGTH OF THE LAGRANGIAN DUAL

To understand the Lagrangian dual problem (LD), we suppose for simplicity that the set X contains a very large but finite number of points $\{x_1, \ldots, x_T\}$. Now

$$
\begin{aligned}
w_{LD} &= \min_{u \geq 0} z(u) \\
&= \min_{u \geq 0}\{\max_{x \in X}[cx + u(d - Dx)]\} \\
&= \min_{u \geq 0}\{\max_{t=1,\ldots,T}[cx^t + u(d - Dx^t)]\} \\
&= \begin{cases} \min \eta \\ \eta \geq cx^t + u(d - Dx^t) \text{ for all } t \\ u \in R_+^T, \eta \in R^1, \end{cases}
\end{aligned}
$$

where the new variable η has been introduced to represent an upper bound on $z(u)$. The latter problem is a linear program. Taking its dual gives:

$$
\begin{aligned}
w_{LD} &= \max \sum_{t=1}^{T} \mu_t (cx^t) \\
&\sum_{t=1}^{T} \mu_t (Dx^t - d) \leq 0 \\
&\sum_{t=1}^{T} \mu_t = 1 \\
&\mu \in R_+^T.
\end{aligned}
$$

Now setting $x = \sum_{t=1}^{T} \mu_t x^t$, with $\sum_{t=1}^{T} \mu_t = 1, \mu \in R_+^T$, we get:

$$
\begin{aligned}
w_{LD} &= \max cx \\
Dx &\leq d \\
x &\in \operatorname{conv}(X).
\end{aligned}
$$

More generally it can be shown that the result still holds when X is the feasible region of any integer program $X = \{x \in Z_+^n : Ax \leq b\}$.

Theorem 10.3 $w_{LD} = \max\{cx : Dx \leq d, x \in \operatorname{conv}(X)\}$.

This theorem tells us precisely how strong a bound we obtain from dualization. In certain cases it is no stronger that the linear programming relaxation.

Corollary If $X = \{x \in Z_+^n : Ax \leq b\}$ and $\mathrm{conv}(X) = \{x \in R_+^n : Ax \leq b\}$, then $w_{LD} = \max\{cx : Ax \leq b, Dx \leq d, x \in R_+^n\}$.

For $STSP$, as constraints (10.6)–(10.8) with $x \geq 0$ describe the convex hull of the incidence vectors of 1-trees, this corollary tells us that w_{LD} gives us precisely the value of the linear programming relaxation of formulation (10.1)–(10.4). This is very interesting because it means that we have found a way to solve a linear program with an exponential number of constraints without treating them explicitly.

The proof of Theorem 10.3 also tells us a good deal about the structure of the Lagrangian dual. It shows that the problem has been convexified so that it becomes a linear program. We also see that

$$w_{LD} = min_{u \geq 0}\{max_{t=1,\ldots,T}[cx^t + u(d - Dx^t)]\}.$$

Thus the Lagrangian dual problem can also be viewed as the problem of minimizing the piecewise linear convex, but nondifferentiable function $z(u)$ (see Figure 10.2).

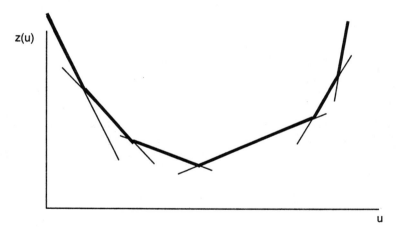

Fig. 10.2 Form of the dual problem

10.3 SOLVING THE LAGRANGIAN DUAL

The linear programming formulation appearing in the proof of Theorem 10.3 provides one way to calculate w_{LD}, though the large number of constraints means that a constraint generation (or cutting plane) approach is required, as described in Chapter 8. An alternative approach that is very simple and

LAGRANGIAN DUALITY

easy to implement without using a linear programming system is a *subgradient algorithm*, which we describe below.

The subgradient algorithm is designed to solve the problem of minimizing a piecewise linear convex function:

$$min_{u \geq 0} f(u), \text{ where } f(u) = max_{t=1,\ldots,T}[a^t u - b_t].$$

In the case of the Lagrangian dual, we have:

$$w_{LD} = \min_{u \geq 0} z(u)$$

where $z(u) = max_{t=1,\ldots,T}[(d - Dx^t)u + cx^t]$.

A subgradient is a straightforward generalization of a gradient.

Definition 10.1 A *subgradient* at u of a convex function $f : R^m \to R^1$ is a vector $\gamma(u) \in R^m$ such that $f(v) \geq f(u) + \gamma(u)^T(v - u)$ for all $v \in R^m$.

For a continuously differentiable convex function f, $\gamma(u) = \nabla f(u) = (\frac{\partial f}{\partial u_1}, \ldots, \frac{\partial f}{\partial u_m})$ is the gradient of f at u.

Subgradient Algorithm for the Lagrangian Dual

Initialization. $u = u^0$.

Iteration k. $u = u^k$.
Solve the Lagrangian problem $IP(u^k)$ with optimal solution $x(u^k)$.
$u^{k+1} = max\{u^k - \mu_k(d - Dx(u^k)), 0\}$
$k \leftarrow k + 1$

The vector $d - Dx(u^k)$ is easily shown to be a subgradient of $z(u)$ at u^k.

The simplicity of this algorithm is amazing. At each iteration one takes a step from the present point u^k in the direction opposite to a subgradient. The difficulty is in choosing the *step lengths* $\{\mu_k\}_{k=1}^{\infty}$.

Theorem 10.4 (a) *If $\sum_k \mu_k \to \infty$, and $\mu_k \to 0$ as $k \to \infty$, then $z(u_k) \to w_{LD}$ the optimal value of LD.*
(b) *If $\mu_k = \mu_0 \rho^k$ for some parameter $\rho < 1$, then $z(u_k) \to w_{LD}$ if μ_0 and ρ are sufficiently large.*
(c) *If $\overline{w} \geq w_{LD}$ and $\mu_k = \epsilon_k[z(u^k) - \overline{w}]/||d - Dx(u^k)||^2$ with $0 < \epsilon_k < 2$, then $z(u^k) \to \overline{w}$, or the algorithm finds u^k with $\overline{w} \geq z(u^k) \geq w_{LD}$ for some finite k.*

This theorem tells us that rule (a) guarantees convergence, but as the series $\{\mu_k\}$ must be divergent (for example $\mu_k = 1/k$), convergence is too slow to be of real practical interest.

On the other hand, step sizes (b) or (c) lead to much faster convergence, but each time with a possible inconvenience.

Using rule (b), the initial values of μ_0 and ρ must be sufficiently large, otherwise the geometric series $\mu_0 \rho^k$ tends to zero too rapidly, and the sequence u^k converges before reaching an optimal point. In practice, rather than decreasing μ_k at each iteration, a geometric decrease is achieved by halving the value of μ_k every ν iterations, where ν is some natural problem parameter such as the number of variables.

Using rule (c), the difficulty is that a dual upper bound $\overline{w} \geq w_{LD}$ is typically unknown. It is more likely in practice that a good primal lower bound $\underline{w} \leq w_{LD}$ is known. Such a lower bound $\underline{w}$ is then used initially in place of $\overline{w}$. However, if $\underline{w} < w_{LD}$, the term $z(u^k) - \underline{w}$ in the numerator of the expression for μ_k will not tend to zero, and so the sequences $\{u^k\}, \{z(u^k)\}$ will not converge. If such behavior is observed, the value of $\underline{w}$ must be increased.

For the symmetric traveling salesman problem, care must be taken because here we are minimizing and the dualized constraints are the equality constraints

$$\sum_{e \in \delta(i)} x_e = 2 \text{ for all } i \in V.$$

The step direction is given by the rule

$$u_i^{k+1} = u_i^k + \mu_k(2 - \sum_{e \in \delta(i)} x_e(u^k)).$$

Here the ith coordinate of the subgradient $2 - \sum_{e \in \delta(i)} x_e(u^k)$ is two minus the number of arcs incident to node i in the optimal 1-tree. The step size, using rule (c), becomes

$$\mu_k = \epsilon_k(\underline{w} - z(u^k))/\sum_{i \in V}(2 - \sum_{e \in \delta(i)} x_e(u^k))^2$$

where a lower bound $\underline{w}$ on w_{LD} is appropriate because $STSP$ is a minimization problem and LD a maximization problem.

Example 10.2 (cont) Suppose that we have found the tour $(1, 2, 3, 4, 5, 1)$ of length 148 by some heuristic, but we have no lower bounds. We update the dual variables using rule (c) from Theorem 10.4. We take $\epsilon_k = 1$ and, as no lower bound is available, we use the value $\overline{w} = 148$ of the tour.

Iteration 1. $u^1 = (0, 0, 0, 0, 0)$. The revised cost matrix is $\overline{c}^1 = c$. An optimal 1-tree is shown in Figure 10.3, and $z(u^1) = 130 \leq z$. As $(2 - \sum_{e \in \delta(i)} x_e(u^k)) = (0, 0, -2, 1, 1)$, we have

$$u^2 = u^1 + [(148 - 130)/6](0, 0, -2, 1, 1).$$

176 LAGRANGIAN DUALITY

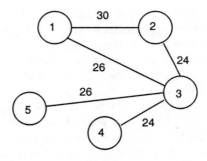

Fig. 10.3 Optimal 1-tree for $\bar{c}^1$

Iteration 2. $u^2 = (0, 0, -6, 3, 3)$. The new cost matrix is

$$(\bar{c}_e^2) = \begin{pmatrix} - & 30 & 32 & 47 & 37 \\ - & - & 30 & 37 & 47 \\ - & - & - & 27 & 29 \\ - & - & - & - & 24 \\ - & - & - & - & - \end{pmatrix}.$$

We obtain $z(u^2) = 143 + 2\sum_i u_i^2 = 143$, and

$$u^3 = u^2 + ((148 - 143)/2)(0, 0, -1, 0, 1).$$

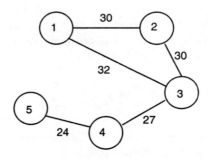

Fig. 10.4 Optimal 1-tree for $\bar{c}^2$

The new optimal 1-tree is shown in Figure 10.4.

Iteration 3. $u^3 = (0, 0, -17/2, 3, 11/2)$.
The new cost matrix is

$$(\bar{c}_e^3) = \begin{pmatrix} - & 30 & 34.5 & 47 & 34.5 \\ - & - & 32.5 & 37 & 44.5 \\ - & - & - & 29.5 & 29 \\ - & - & - & - & 21.5 \\ - & - & - & - & - \end{pmatrix}.$$

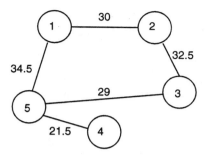

Fig. 10.5 Optimal 1-tree for $\bar{c}^3$

The new optimal 1-tree is shown in Figure 10.5 and we obtain the lower bound $z(u^3) = 147.5$. As the cost data c are integral, we know that z is integer valued and so $z \geq \lceil 147.5 \rceil = 148$. As a solution of cost 148 is known, the corresponding solution has been proved optimal. ∎

As the subgradient algorithm is often terminated before the optimal value w_{LD} is attained, and also as there is in most cases a duality gap ($w_{LD} > z$), Lagrangian relaxation must typically be embedded in a branch-and-bound algorithm.

10.4 LAGRANGIAN HEURISTICS AND VARIABLE FIXING

Once the dual variables u begin to approach the set of optimal solutions, a solution $x(u)$ is obtained that is hopefully "close" to being primal feasible every time that a Lagrangian subproblem $IP(u)$ is solved. In the $STSP$ many nodes of the 1-tree will have degree 2, and so the solution is not far from being a tour, while for the UFL, many clients are served exactly once, and only a few are not served at all. Therefore it is often straightforward to devise a simple heuristic that converts $x(u)$ into a feasible solution without greatly decreasing/increasing its value/cost. Below we examine this simple idea for set covering problems, as well as the possibility of fixing some variables once good primal and dual solutions are available.

Consider an instance of the set-covering problem

$$\min\{\sum_{j \in N} c_j x_j : \sum_{j \in N} a_{ij} x_j \geq 1 \text{ for } i \in M, x \in B^n\},$$

with $a_{ij} \in \{0, 1\}$ for $i \in M, j \in N$. The Lagrangian relaxation in which all the covering constraints are dualized is

$$z(u) = \sum_{i \in M} u_i + \min\{\sum_{j \in N} (c_j - \sum_{i \in M} u_i a_{ij}) x_j : x \in B^n\}$$

for $u \geq 0$.

One simple possibility is to take an optimal solution $x(u)$ of this relaxation, drop all rows covered by the solution $x(u)$, that is, the rows $i \in M$ for which $\sum_{j \in N} a_{ij} x_j(u) \geq 1$, and solve the remaining smaller covering problem by a greedy heuristic. If y^* is the heuristic solution, then $x^H = x(u) + y^*$ is a feasible solution. It is then worth checking whether it cannot be improved by setting to zero some of the variables with $x_j(u) = 1$.

Once a heuristic solution has been found, it is also possible to use the Lagrangian for variable fixing. If $\bar{z}$ is the incumbent value, then any better feasible solution x satisfies $\sum_{i \in M} u_i + \min \sum_{j \in N} (c_j - \sum_{i \in M} u_i a_{ij}) x_j \leq cx < \bar{z}$. Let $N_1 = \{j \in N : c_j - \sum_{i \in M} u_i a_{ij} > 0\}$ and $N_0 = \{j \in N : c_j - \sum_{i \in M} u_i a_{ij} < 0\}$.

Proposition 10.5 If $k \in N_1$ and $\sum_{i \in M} u_i + \sum_{j \in N_0} (c_j - \sum_{i \in M} u_i a_{ij}) + (c_k - \sum_{i \in M} u_i a_{ik}) \geq \bar{z}$, then $x_k = 0$ in any better feasible solution.

If $k \in N_0$ and $\sum_{i \in M} u_i + \sum_{j \in N_0 \setminus \{k\}} (c_j - \sum_{i \in M} u_i a_{ij}) \geq \bar{z}$, then $x_k = 1$ in any better feasible solution.

Example 10.3 Consider a set-covering instance with $m = 4, n = 6$,

$$c = (6, 6, 11, 5, 8, 8) \text{ and } a_{ij} = \begin{pmatrix} 1 & 0 & 1 & 0 & 0 & 1 \\ 0 & 1 & 0 & 1 & 1 & 0 \\ 1 & 1 & 1 & 0 & 0 & 0 \\ 0 & 0 & 1 & 0 & 1 & 0 \end{pmatrix}.$$

Taking $u = (4, 4, 3, 3)$, the Lagrangian subproblem $IP(u)$ takes the form

$$z(u) = 14 + \min\{-1x_1 + 0x_2 + 1x_3 + 1x_4 + 1x_5 + 1x_6 : x \in B^6\}.$$

An optimal solution is clearly $x(u) = (1, 0, 0, 0, 0, 0)$ wth $z(u) = 13$. The solution $x(u)$ covers rows 1 and 3, so the remaining problem to be solved heuristically is

$$\begin{array}{rrrrrrl}
\min 6x_1 + & 6x_2 + & 11x_3 + & 5x_4 + & 8x_5 + & 8x_6 & \\
& x_2 & & x_4 + & x_5 & & \geq 1 \\
& & x_3 & & x_5 & & \geq 1 \\
& & & x \in & B^6 & &
\end{array}$$

for which a greedy heuristic gives the solution $y^* = (0, 0, 0, 0, 1, 0)$. Adding together these two vectors, we obtain the heuristic solution $x^H = (1, 0, 0, 0, 1, 0)$ with cost 14. Thus we now know that the optimal value lies between 13 and 14.

Now using Proposition 10.5, we see that with $N_0 = \{1\}$ and $N_1 = \{3, 4, 5, 6\}$, $x_1 = 1$ and $x_3 = x_4 = x_5 = x_6 = 0$ in any solution whose value is less than 14. ∎

10.5 CHOOSING A LAGRANGIAN DUAL

Suppose the problem to be solved is of the form:

(IP)
$$z = \max cx \\ A^1 x \le b^1 \\ A^2 x \le b^2 \\ x \in Z_+^n.$$

If one wishes to tackle the problem by Lagrangian relaxation, there is a choice to be made. Should one dualize one or both sets of constraints, and if so, which sets? The answer must be based on a trade-off between

(i) the *strength* of the resulting Lagrangian dual bound w_{LD},

(ii) *ease of solution* of the Lagrangian subproblems $IP(u)$, and

(iii) *ease of solution* of the Lagrangian dual problem: $w_{LD} = \min_{u \ge 0} z(u)$.

Concerning (i), Theorem 10.3 gives us precise information about the strength of the bound.

Concerning (ii), the ease of solution of $IP(u)$ is problem specific. However, we know that if $IP(u)$ is "easy" in the sense of reducing to a linear program, that is $IP(u)$ involves maximization over $X = \{x \in Z_+^n : Ax \le b\}$ and $\operatorname{conv}(X) = \{x \in R_+^n : Ax \le b\}$, then solving the linear programming relaxation of IP is an alternative to Lagrangian relaxation.

Concerning (iii), the difficulty using the subgradient (or other) algorithms is hard to estimate a priori, but the number of dual variables is at least some measure of the probable difficulty.

To demonstrate these trade-offs, consider the *Generalized Assignment Problem (GAP)*:

$$z = \max \sum_{j=1}^n \sum_{i=1}^m c_{ij} x_{ij} \\ \sum_{j=1}^n x_{ij} \le 1 \text{ for } i = 1, \ldots, m \\ \sum_{i=1}^m a_{ij} x_{ij} \le b_j \text{ for } j = 1, \ldots, n \\ x \in B^{mn}.$$

We consider three possible Lagrangian relaxations. In the first we dualize both sets of constraints giving $w_{LD}^1 = \min_{u \ge 0, v \ge 0} w^1(u, v)$ where

$$w^1(u,v) = \max_x \sum_{j=1}^n \sum_{i=1}^m (c_{ij} - u_i - a_{ij} v_j) x_{ij} + \sum_{i=1}^m u_i + \sum_{j=1}^n v_j b_j \\ x \in B^{mn}.$$

Here we dualize the first set of assignment constraints giving $w_{LD}^2 = \min_{u \ge 0} w^2(u)$ where

$$w^2(u) = \max_x \sum_{j=1}^n \sum_{i=1}^m (c_{ij} - u_i)x_{ij} + \sum_{i=1}^m u_i$$
$$\sum_{i=1}^m a_{ij}x_{ij} \leq b_j \text{ for } j = 1,\ldots,n$$
$$x \in B^{mn},$$

and here we dualize the knapsack constraints giving $w_{LD}^3 = \min_{v \geq 0} w^3(v)$ where

$$w^3(v) = \max_x \sum_{j=1}^n \sum_{i=1}^m (c_{ij} - a_{ij}v_j)x_{ij} + \sum_{j=1}^n v_j b_j$$
$$\sum_{j=1}^n x_{ij} \leq 1 \text{ for } i = 1,\ldots,m$$
$$x \in B^{mn}.$$

Based on Theorem 10.3, we know that $w_{LD}^1 = w_{LD}^3 = z_{LP}$ as for each i, conv$\{x : \sum_{j=1}^n x_{ij} \leq 1, x_{ij} \in \{0,1\} \text{ for } j = 1,\ldots,n\} = \{x : \sum_{j=1}^n x_{ij} \leq 1, 0 \leq x_{ij} \leq 1 \text{ for } j = 1,\ldots,n\}$. The values of $w^1(u,v)$ and $w^3(v)$ can both be calculated by inspection. To calculate $w^1(u,v)$, note that the problem decomposes variable by variable, while for $w^3(u,v)$ the problem decomposes into a simple problem for each $j = 1,\ldots,n$. In terms of solving the Lagrangian dual problems, calculating w_{LD}^3 appears easier than calculating w_{LD}^1 because there are only m as opposed to $m + n$ dual variables.

The second relaxation potentially gives a tighter bound $w_{LD}^2 \leq z_{LP}$ as in general for fixed j, conv$\{x : \sum_{i=1}^m a_{ij}x_{ij} \leq b_j, x_{ij} \in \{0,1\}^m\} \subset \{x : \sum_{i=1}^m a_{ij}x_{ij} \leq b_j, 0 \leq x_{ij} \leq 1 \text{ for } i = 1,\ldots,m\}$. However, here the Lagrangian subproblem involves the solution of m 0–1 knapsack problems.

10.6 NOTES

10.1 Many of the properties of the Lagrangian dual can be found in [Eve63]. The successful solution of what were at the time very large *TSP*s [HelKar70], [HelKar71] made the approach popular.

10.2 The application to integer programming and in particular Theorem 10.3 and its consequences were explored in [Geo74].

10.3 The use of the subgradient algorithm to solve the Lagrangian dual again stems from [HelKar70]. Detailed studies and analysis of the subgradient approach can be found in [HelWolCro74] and [Gof77]. Recently more sophisticated nondifferentiable optimization techniques have been used; see [Lemetal95]. Simple multiplier adjustment methods have also been tried for various problems, among them the uncapacitated facility location problem [Erl78]. An alternative approach, Dantzig-Wolfe decomposition, is treated in the next chapter.

10.5 A comparison of different Lagrangian relaxations for the capacitated facility location problem can be found in [CorSriThi91]. By duplicating variables, dualizing the equations identifying variables, and solving separate subproblems for each set of distinct variables, Lagrangian relaxation can be used to get stronger bounds in certain cases; see Exercise 10.6. Lagrangian decomposition is one of several names given to this idea [JorNas86], [GuiKim87].

Lagrangian relaxation is an important practical tool for many structured problems. Surveys on the applications of Lagrangian duality include [Fis81] and [Beas93]. It suffices to open journals such as *Operations Research, Management Science* or the *European Journal of Operations Research* to find a wide variety of applications. [Beas93] lists 21 applications based on Lagrangian relaxation that were found in these three journals just for 1991.

10.7 EXERCISES

1. Consider an instance of UFL with $m = 6, n = 5$, delivery costs

$$c_{ij} = \begin{pmatrix} 6 & 2 & 1 & 3 & 5 \\ 4 & 10 & 2 & 6 & 1 \\ 3 & 2 & 4 & 1 & 3 \\ 2 & 0 & 4 & 1 & 4 \\ 1 & 8 & 6 & 2 & 5 \\ 3 & 2 & 4 & 8 & 1 \end{pmatrix}$$

and fixed costs $f = (4, 8, 11, 7, 5)$. Using the dual vector $u = (5, 6, 3, 2, 6, 4)$, solve the Lagrangian subproblem $IP(u)$ to get an optimal solution $(x(u), y(u))$ and lower bound $z(u)$. Modify the dual solution $(x(u), y(u))$ to construct a good primal feasible solution. How far is this solution from optimal?

2. Suppose one dualizes the constraints $x_{ij} \leq y_j$ in the strong formulation of UFL. How strong is the resulting Lagrangian dual bound, and how easy is the solution of the Lagrangian subproblem?

3. Use Lagrangian relaxation to solve the $STSP$ instance with distances

$$(c_e) = \begin{pmatrix} - & 8 & 2 & 14 & 26 & 13 \\ - & - & 7 & 4 & 16 & 8 \\ - & - & - & 23 & 14 & 9 \\ - & - & - & - & 12 & 6 \\ - & - & - & - & - & 5 \end{pmatrix}.$$

4. Use Lagrangian relaxation to solve the GAP instance with

$$(c_{ij}) = \begin{pmatrix} 6 & 10 & 1 \\ 12 & 12 & 5 \\ 15 & 4 & 3 \\ 10 & 3 & 9 \\ 8 & 9 & 5 \end{pmatrix}, (a_{ij}) = \begin{pmatrix} 5 & 7 & 2 \\ 14 & 8 & 7 \\ 10 & 6 & 12 \\ 8 & 4 & 15 \\ 6 & 12 & 5 \end{pmatrix} \text{ and } b = (15, 15, 15)^T.$$

5. Consider a 0–1 knapsack problem

$$z = \max 10y_1 + 4y_2 + 14y_3$$
$$3y_1 + y_2 + 4y_3 \le 4$$
$$y \in B^3.$$

Construct a Lagrangian dual by dualizing the knapsack constraint. What is the optimal value of the dual variable?

Suppose one runs the subgradient algorithm using step size (b) in Theorem 10.4, starting with $u^0 = 0, \mu_0 = 1$ and $\rho = \frac{1}{2}$. Show that the subgradient algorithm does not reach the optimal dual solution.

6. *Lagrangian Decomposition.* Consider the problem $z = \max\{cx : A^i x \le b^i \text{ for } i = 1, 2, x \in Z_+^n\}$ with the reformulation

$$\max\{\alpha c x^1 + (1-\alpha) c x^2 : A^i x^i \le b^i \text{ for } i = 1, 2, x^1 - x^2 = 0, x^i \in Z_+^n \text{ for } i = 1, 2\}$$

for $0 < \alpha < 1$.

Consider the Lagrangian dual of this formulation in which the n constraints $x^1 - x^2 = 0$ are dualized. What is the strength of this dual?

7. Consider the capacitated facility location problem

$$\min \sum_{i \in M} \sum_{j \in N} c_{ij} x_{ij} + \sum_{j \in N} f_j y_j$$
$$\sum_{j \in N} x_{ij} = a_i \text{ for } i \in M$$
$$\sum_{i \in M} x_{ij} \le b_j y_j \text{ for } j \in N$$
$$x_{ij} \le \min\{a_i, b_j\} y_j \text{ for } i \in M, j \in N$$
$$\sum_{j \in N} b_j y_j \ge \sum_{i \in M} a_i$$
$$x \in R_+^{mn}, y \in B^n.$$

Discuss the advantages and disadvantages of different Lagrangian relaxations. Which would you choose to implement and why?

8. Consider the assignment problem with budget constraint

$$\max \sum_{i \in M} \sum_{j \in N} c_{ij} x_{ij}$$
$$\sum_{j \in N} x_{ij} = 1 \text{ for } i \in M$$
$$\sum_{i \in M} x_{ij} = 1 \text{ for } j \in N$$
$$\sum_{i \in M} \sum_{j \in N} a_{ij} x_{ij} \leq b$$
$$x \in B^{mn}.$$

Discuss the strength of different possible Lagrangian relaxations, and the ease or difficulty of solving the Lagrangian subproblems, and the Lagrangian dual.

11
Column Generation Algorithms

11.1 INTRODUCTION

One of the recurring ideas in optimization and in this text is that of decomposition. We have several times considered what happens when the integer programming problem *(IP)* $\max\{cx : x \in X\}$ has a feasible region X that can be written as the intersection of two or more sets with structure $X = \cap_{k=0}^{K} X^k$ for some $K \geq 1$. Even more particular is the case where the constraints take the form:

$$\begin{array}{rcl} A^1 x^1 \;\; +A^2 x^2 \;+\ldots\; +A^K x^K & = & b \\ D^1 x^1 & \leq & d_1 \\ \ldots & \leq & \cdot \\ \ldots & \leq & \cdot \\ D^K x^K & \leq & d_K \\ x^1 \in Z_+^{n_1}, \;\; \ldots \;\; \ldots \;\; , x^K \in Z_+^{n_K} & & \end{array}$$

(coupling constraints ← ; individual constraints, sets X^k ←)

so that the sets $X^k = \{x^k \in Z_+^{n_k} : D^k x^k \leq d_k\}$ are independent for $k = 1, \ldots, K$, and only the *joint* constraints $\sum_{k=1}^{K} A^k x^k = b$ link together the different sets of variables.

Given an objective function $\max \sum_{k=1}^{K} c^k x^k$, two earlier approaches that would permit us to benefit from such structure are cut generation, in which we would try to generate valid inequalities for each subset $X^k, k = 1, \ldots, K$, and Lagrangian relaxation, in which we would dualize the joint constraints so as to obtain a dual problem:

$$\min_{u} L(u),$$

185

where $L(u) = \max\{\sum_{k=1}^{K} c^k x^k + u^T(b - \sum_{k=1}^{K} A^k x^k)\}$

$$L(u) = \max\{\sum_{k=1}^{K}(c^k - uA^k)x^k + ub : x^k \in X^k \text{ for } k = 1,\ldots,K\},$$

and the calculation of $L(u)$ breaks up into K distinct subproblems:

$$L(u) = \left[\sum_{k=1}^{K} \max\{(c^k - uA^k)x^k : x^k \in X^k\}\right] + ub.$$

In this chapter we examine a third way to exploit the structure of integer programs of the above form. Throughout we assume that each of the sets X^k is bounded for $k = 1,\ldots,K$. The approach essentially involves solving an equivalent problem of the form:

$$\max\{\sum_{k=1}^{K} \gamma^k \lambda^k : \sum_{k=1}^{K} B^k \lambda^k = \beta, \lambda^k \geq 0 \text{ integer for } k = 1,\ldots,K\}$$

where each matrix B^k has a very large number of columns, one for each of the feasible points in X^k, and each vector λ^k contains the corresponding variables.

UFL: For example, we now derive an alternative formulation of this type for the uncapacitated facility location problem UFL. Here the locations $j = 1,\ldots,n$ correspond to the indices $k = 1,\ldots,K$. For each nonempty subset $S \subseteq M$ of clients, let $\lambda_S^j = 1$ if depot j satisfies the demand of client set S. This then leads to the formulation:

$$\min \sum_{j \in N} \sum_{S \neq \phi} (\sum_{i \in S} c_{ij} + f_j)\lambda_S^j \quad (11.1)$$

$$\sum_{j \in N} \sum_{S \neq \phi, i \in S} \lambda_S^j = 1 \text{ for } i \in M \quad \text{(each client } i \text{ is served)} \quad (11.2)$$

$$\sum_{S \neq \phi} \lambda_S^j \leq 1 \text{ for } j \in N \quad (11.3)$$

$$\lambda_S^j \in \{0,1\} \text{ for } \phi \neq S \subseteq M, j \in N. \quad (11.4)$$

Here the cost of λ_S^j is the cost of opening depot j and serving the clients in S from depot j. The first constraints again impose that each client is served, while the second set of constraints ensure that at most one subset of clients is assigned to depot j. In practice the latter constraints are typically unnecessary. Why? (Since it is uncapacitated FL, this holds)

Thus the problems we wish to solve here are integer programs with an enormous number of variables, where the columns are often described implicitly as the incidence vectors of certain subsets of a set, that is tours, client subsets, and so on. Below we show how such (large) formulations of an integer program, called *Master Problems*, also arise by reformulation. We then consider how to solve the linear programming relaxation of these Master Problems, and relate the strength of this relaxation to that obtained by Lagrangian duality, or by the use of cutting planes. Finally we consider what to do when the linear programming solution is not integral, and we must resort to enumeration, leading to *IP Column Generation* or *Branch-and-Price* algorithms.

11.2 DANTZIG-WOLFE REFORMULATION OF AN IP

Consider the problem in the form: → *coupling constraint*

$$(IP) \quad z = \max\{\sum_{k=1}^{K} c^k x^k : \sum_{k=1}^{K} A^k x^k = b, x^k \in X^k \text{ for } k = 1, \ldots, K\} \quad (11.5)$$

where $X^k = \{x^k \in Z_+^{n_k} : D^k x^k \leq d_k\}$ for $k = 1, \ldots, K$. Assuming that each set X^k contains a large but finite set of points $\{x^{k,t}\}_{t=1}^{T_k}$, we have that $X^k = \{x^k \in R^{n_k} : x^k = \sum_{t=1}^{T_k} \lambda_{k,t} x^{k,t}, \sum_{t=1}^{T_k} \lambda_{k,t} = 1, \lambda_{k,t} \in \{0,1\} \text{ for } t = 1, \ldots, T_k\}$.

Now substituting for x^k leads to an equivalent *IP Master Problem*:

$$(IPM) \quad \begin{aligned} & z = \max \sum_{k=1}^{K} \sum_{t=1}^{T_k} (c^k x^{k,t}) \lambda_{k,t} \\ & \sum_{k=1}^{K} \sum_{t=1}^{T_k} (A^k x^{k,t}) \lambda_{k,t} = b \\ & \sum_{t=1}^{T_k} \lambda_{k,t} = 1 \text{ for } k = 1, \ldots, K \\ & \lambda_{k,t} \in \{0,1\} \text{ for } t = 1, \ldots, T_k \text{ and } k = 1, \ldots, K. \end{aligned}$$

Continuing with problem UFL, suppose we start from the weak formulation

$$\min \sum_{i \in M} \sum_{j \in N} c_{ij} x_{ij} + \sum_{j \in N} f_j y_j \quad (11.6)$$
$$\sum_{j \in N} x_{ij} = 1 \text{ for } i \in M \quad (11.7)$$
$$\sum_{i \in M} x_{ij} \leq m y_j \text{ for } j \in N \quad (11.8)$$
$$x \in B^{|M| \times |N|}, y \in B^{|N|}. \quad (11.9)$$

Here we can take (11.7) as the joint constraints, and $X^k = \{(x_{1k}, \ldots, x_{mk}, y_k) : \sum_{i \in M} x_{ik} \leq m y_k, x_{ik} \in B^1 \text{ for } i \in M, y_k \in \{0,1\}\}$. The points in X^k are $\{(x_S^k, 1)\}_{S \subseteq M}$, where x_S^k is the incidence vector of $S \subseteq M$, and $(\mathbf{0}, 0)$ with associated variables λ_S^k, ν^k respectively leading to the *IP* Master Problem:

$$\min \sum_{j \in N} [\sum_{S \neq \phi} (\sum_{i \in S} c_{ij} + f_j) \lambda_S^j + f_j \lambda_\emptyset^j]$$
$$\sum_{j \in N} \sum_{S \neq \phi, i \in S} \lambda_S^j = 1 \text{ for } i \in M$$
$$\sum_{S \neq \phi} \lambda_S^j + \lambda_\emptyset^j + \nu^j = 1 \text{ for } j \in N$$
$$\lambda_S^j \in \{0,1\} \text{ for } S \subseteq M, j \in N, \nu^j \in \{0,1\} \text{ for } j \in N.$$

Observe that as $f_j \geq 0$, variable $\lambda_\emptyset^j$ is dominated by ν^j and can be dropped. Now this formulation and (11.1)–(11.4) are identical if we take ν^j to be the slack variable in (11.3).

11.3 SOLVING THE MASTER LINEAR PROGRAM

Here we use a column generation algorithm to solve the linear programming relaxation of the Integer Programming Master Problem, called the *Linear Programming Master Problem*:

$$(LPM) \quad \begin{aligned} z^{LPM} &= \max \sum_{k=1}^{K} \sum_{t=1}^{T_k} (c^k x^{k,t}) \lambda_{k,t} \\ & \sum_{k=1}^{K} \sum_{t=1}^{T_k} (A^k x^{k,t}) \lambda_{k,t} = b \\ & \sum_{t=1}^{T_k} \lambda_{k,t} = 1 \text{ for } k = 1, \ldots, K \\ & \lambda_{k,t} \geq 0 \text{ for } t = 1, \ldots, T^k, k = 1, \ldots, K \end{aligned}$$

where there is a column $\begin{pmatrix} c^k x \\ A^k x \\ e_k \end{pmatrix}$ for each $x \in X^k$. Below we will use $\{\pi_i\}_{i=1}^m$ as the dual variables associated with the joint constraints, and $\{\mu_k\}_{k=1}^K$ as dual variables for the second set of constraints, known as *convexity* constraints.

The idea is to solve the linear program by the primal simplex algorithm. However, the pricing step of choosing a column to enter the basis must be modified because of the enormous number of columns. Rather than pricing the columns one by one, the problem of finding a column with the largest reduced price is itself a set of K optimization problems.

Initialization. We suppose that a subset of columns (at least one for each k) is available, providing a feasible *Restricted Linear Programming Master Problem*

$$(RLPM) \quad \begin{aligned} \tilde{z}^{LPM} &= \max \tilde{c} \tilde{\lambda} \\ \tilde{A} \tilde{\lambda} &= \tilde{b} \\ \tilde{\lambda} &\geq 0 \end{aligned}$$

where $\tilde{b} = \begin{pmatrix} b \\ 1 \end{pmatrix}$, $\tilde{A}$ is generated by the available set of columns and is a submatrix of

$$\begin{pmatrix} A^1 x^{1,1} & .. & A^1 x^{1,T_1} & A^2 x^{2,1} & .. & A^2 x^{2,T_2} & & A^K x^{K,1} & .. & A^K x^{K,T_K} \\ 1 & .. & 1 & & & & & & & \\ & & & 1 & .. & 1 & & & & \\ & & & & & & .. & & & \\ & & & & & & & & 1 & .. & 1 \end{pmatrix},$$

and $\tilde{c}, \tilde{\lambda}$ are the corresponding costs and variables. Solving $RLPM$ gives an optimal primal solution $\tilde{\lambda}^*$ and an optimal dual solution $(\pi, \mu) \in R^m \times R^K$.

Primal Feasibility. Any feasible solution of $RLPM$ is feasible for LPM. In particular, $\tilde{\lambda}^*$ is a feasible solution of LPM, and so $\tilde{z}^{LPM} = \tilde{c}\tilde{\lambda}^* = \sum_{i=1}^{m} \pi_i b_i + \sum_{k=1}^{K} \mu_k \leq z^{LPM}$.

Optimality Check for LPM. We need to check whether (π, μ) is dual feasible for LPM. This involves checking for each column, that is for each k, and for each $x \in X^k$ whether the reduced price $c^k x - \pi A^k x - \mu_k \leq 0$. Rather than examining each point separately, we treat all points in X^k implicitly by solving an optimization subproblem:

$$\zeta_k = \max\{(c^k - \pi A^k)x - \mu_k : x \in X^k\}.$$

Stopping Criterion. If $\zeta_k = 0$ for $k = 1, \ldots, K$, the solution (π, μ) is dual feasible for LPM, and so $z^{LPM} \leq \sum_{i=1}^{m} \pi_i b_i + \sum_{k=1}^{K} \mu_k$. As the value of the primal feasible solution $\tilde{\lambda}$ equals that of this upper bound, $\tilde{\lambda}$ is optimal for LPM.

Generating a New Column. If $\zeta_k > 0$ for some k, the column corresponding to the optimal solution $\tilde{x}^k$ of the subproblem has positive reduced price. Introducing the column $\begin{pmatrix} c^k \tilde{x}^k \\ A^k \tilde{x}^k \\ e_k \end{pmatrix}$ leads to a new Restricted Linear Programming Master Problem that can be easily reoptimized (e.g., by the primal simplex algorithm).

A Dual (Upper) Bound. From the subproblem, we have that $\zeta_k \geq (c^k - \pi A^k)x - \mu_k$ for all $x \in X^k$. It follows that $(c^k - \pi A^k)x - \mu_k - \zeta_k \leq 0$ for all $x \in X^k$. Therefore setting $\zeta = (\zeta_1, \ldots, \zeta_K)$, we have that $(\pi, \mu + \zeta)$ is dual feasible in LPM. Therefore

$$z^{LPM} \leq \pi b + \sum_{k=1}^{K} \mu_k + \sum_{k=1}^{K} \zeta_k.$$

These different observations lead directly to an algorithm for LPM that terminates when $\zeta_k = 0$ for $k = 1, \ldots, K$. However, as in Lagrangian relaxation, it may be possible to terminate earlier.

An Alternative Stopping Criterion. If the subproblem solutions $(\tilde{x}^1, \ldots, \tilde{x}^K)$ satisfy the original joint constraints $\sum_{k=1}^{K} A^k x^k = b$, then $(\tilde{x}^1, \ldots, \tilde{x}^K)$ is optimal.

This follows because $\zeta_k = (c^k - \pi A^k)\tilde{x}^k - \mu_k$ implies that $\sum_k c^k \tilde{x}^k = \sum_k \pi A^k \tilde{x}^k + \sum_k \mu_k + \sum_k \zeta_k = \pi b + \sum_k \mu_k + \sum_k \zeta_k$. Therefore the primal feasible solution has the same value as the upper bound on z^{LPM}.

11.3.1 STSP by Column Generation

Here we consider the application of the above algorithm to solve the Master Linear Program of a problem in which there is just a single subproblem. We again consider the symmetric traveling salesman problem, which can be written as

$$\min\{\sum_{e \in E} c_e x_e : \sum_{e \in \delta(i)} x_e = 2 \text{ for } i \in N, x \in X^1\}$$

where

$$X^1 = \{x \in Z_+^m : \sum_{e \in \delta(1)} x_e = 2, \sum_{e \in E(S)} x_e \leq |S| - 1 \text{ for } \phi \subset S \subset N \setminus \{1\},$$

$$\sum_{e \in E} x_e = n\}$$

is the set of incidence vectors of 1-trees.

Writing $x_e = \sum_{t: e \in E^t} \lambda_t$, where $G^t = (N, E^t)$ is the tth 1-tree, the degree constraints become $\sum_{e \in \delta(i)} x_e = \sum_{e \in \delta(i)} \sum_{t: e \in E^t} \lambda_t = \sum_t d_i^t \lambda_t = 2$ where d_i^t is the degree of node i in the 1-tree G^t. Thus the corresponding Linear Programming Master is

$$(LPM) \quad \begin{array}{c} \min \sum_{t=1}^{T_1} (cx^t) \lambda_t \\ \sum_{t=1}^{T_1} d_i^t \lambda_t = 2 \text{ for } i \in N \\ \sum_{t=1}^{T_1} \lambda_t = 1 \\ \lambda \in R_+^T \end{array}$$

with which we associate dual variables $\{u_i\}_{i=1}^n$ to the degree constraints, and dual variable μ to the convexity constraint. The corresponding single subproblem is

$$\zeta_1 = \min\{\sum_{e \in E} (c_e - u_i - u_j) x_e - \mu : x \in X^1\}$$

as the 1-tree G^t has reduced cost $cx^t - \sum_{i \in N} d_i^t u_i - \mu = cx^t - \sum_{i \in N} u_i \sum_{e \in \delta(i)} x_e^t - \mu = \sum_{e \in E} (c_e - u_i - u_j) x_e^t - \mu$, where x_e^t for $e \in E$ are the edge variables of the 1-tree G^t, and $e = (i, j)$ for $e \in E$.

Note that because we are dealing with 1-trees, $d_1^t = 2$ for all t, and so the first equation in LPM is twice the convexity constraint. As a result we can drop the convexity constraint.

Example 11.1 Consider an instance of $STSP$ with distance matrix

$$c_e = \begin{pmatrix} . & 7 & 2 & 1 & 5 \\ & . & 3 & 6 & 8 \\ & & . & 4 & 2 \\ & & & . & 9 \\ & & & & . \end{pmatrix}$$

We initialize with a restricted LPM having 7 columns, corresponding to a tour of length 28 and six 1-trees chosen arbitrarily

$$\begin{array}{rrrrrrrrl}
\min & 28\lambda_1 & +25\lambda_2 & +21\lambda_3 & +19\lambda_4 & +22\lambda_5 & +18\lambda_6 & +28\lambda_7 & \\
& 2\lambda_1 & +2\lambda_2 & +2\lambda_3 & +2\lambda_4 & +2\lambda_5 & +2\lambda_6 & +2\lambda_7 & = 2 \\
& 2\lambda_1 & +2\lambda_2 & +2\lambda_3 & +1\lambda_4 & +1\lambda_5 & +2\lambda_6 & +3\lambda_7 & = 2 \\
& 2\lambda_1 & +3\lambda_2 & +2\lambda_3 & +3\lambda_4 & +2\lambda_5 & +3\lambda_6 & +1\lambda_7 & = 2 \\
& 2\lambda_1 & +2\lambda_2 & +3\lambda_3 & +3\lambda_4 & +3\lambda_5 & +1\lambda_6 & +1\lambda_7 & = 2 \\
& 2\lambda_1 & +1\lambda_2 & +1\lambda_3 & +1\lambda_4 & +2\lambda_5 & +2\lambda_6 & +3\lambda_7 & = 2 \\
& & & & \lambda & \geq & 0. & &
\end{array}$$

The resulting linear programming solution is $\lambda = (0, 0, \frac{1}{4}, 0, \frac{1}{4}, \frac{1}{4}, \frac{1}{4})$ with cost 22.5 and dual solution $u = (\frac{151}{8}, -1, -\frac{11}{2}, -\frac{5}{4}, 0)$. The corresponding reduced cost matrix for the subproblem is

$$\begin{pmatrix}
\cdot & -\frac{87}{8} & -\frac{91}{8} & -\frac{133}{8} & -\frac{111}{8} \\
 & \cdot & \frac{19}{2} & \frac{33}{4} & 9 \\
 & & \cdot & \frac{43}{4} & \frac{15}{2} \\
 & & & \cdot & \frac{41}{4} \\
 & & & & \cdot
\end{pmatrix}$$

The optimal 1-tree is $x_{14} = x_{15} = x_{24} = x_{25} = x_{35} = 1$ with $\zeta = -\frac{23}{4}$. Therefore $22.5 + \zeta = 16.75 \leq z^{LP} \leq 22.5$.

We start a new iteration by introducing this 1-tree as a new column in the restricted master with cost 22, and degrees (2,2,1,2,3). The new linear programming solution is $\lambda = (0, 0, \frac{1}{3}, 0, 0, \frac{1}{3}, 0, \frac{1}{3})$ with cost 20.333 and dual solution $u = (\frac{65}{6}, \frac{1}{3}, -\frac{5}{3}, \frac{2}{3}, 0)$.

The corresponding reduced cost matrix for the subproblem is

$$\begin{pmatrix}
\cdot & -\frac{25}{6} & -\frac{43}{6} & -\frac{21}{2} & -\frac{35}{6} \\
 & \cdot & \frac{13}{3} & 5 & \frac{23}{3} \\
 & & \cdot & 5 & \frac{11}{3} \\
 & & & \cdot & \frac{25}{3} \\
 & & & & \cdot
\end{pmatrix}$$

The optimal 1-tree is $x_{13} = x_{14} = x_{23} = x_{24} = x_{35} = 1$ with $\zeta = -\frac{14}{3}$. The lower bound of $20.333 - \frac{14}{3} = 15.667$ is not as good as that obtained before. Therefore we now have $16.75 \leq z^{LP} \leq 20.333$.

Again we introduce this 1-tree as a new column in the restricted master with cost 14, and degrees (2,2,3,2,1). The new linear programming solution is $\lambda = (0, 0, 0, 0, 0, 0, 0, \frac{1}{2}, \frac{1}{2})$ with cost 18 and dual solution $u = (13, 0, -4, 0, 0)$.

The corresponding reduced cost matrix for the subproblem is

$$\begin{pmatrix}
\cdot & -6 & -7 & -12 & -8 \\
 & \cdot & 7 & 6 & 8 \\
 & & \cdot & 8 & 6 \\
 & & & \cdot & 9 \\
 & & & & \cdot
\end{pmatrix}$$

The optimal 1-tree is $x_{14} = x_{15} = x_{23} = x_{24} = x_{35} = 1$ with $\zeta = -1$. The lower bound on z^{LPM} increases to $18 - 1 = 17$. As this 1-tree is a tour, it follows from the alternative stopping criterion that it is optimal. Alternatively one can check that its real cost is 17. ∎

11.3.2 Strength of the Linear Programming Master

How strong is the linear programming relaxation of the Master Problem? Is there some hope that it will solve the original problem IP?

Proposition 11.1

$$z^{LPM} = \max\{\sum_{k=1}^{K} c^k x^k : \sum_{k=1}^{K} A^k x^k = b, x^k \in conv(X^k) \text{ for } k = 1, \ldots, K\}.$$

Proof. LPM can be obtained from the original problem IP by substituting $x^k = \sum_{t=1}^{T_k} x^{k,t} \lambda_{k,t}, \sum_{t=1}^{T_k} \lambda_{k,t} = 1, \lambda_{k,t} \geq 0$ for $t = 1, \ldots, T_k$. This is equivalent to substituting $x^k \in \text{conv}(X^k)$. ∎

As discussed in the introduction to this chapter, when IP is decomposable, Lagrangian relaxation and cutting plane algorithms are two possible alternative approaches. Specifically let w_{LD} be the value of the Lagrangian dual when the joint constraints $\sum_{k=1}^{K} A^k x^k = b$ are dualized, and let z^{CUT} be the value obtained when cutting planes are added to the linear programming relaxation of IP using an exact separation algorithm for each of the sets $\text{conv}(X^k)$ for $k = 1, \ldots, K$.

The next result, showing that all three approaches are in some sense equivalent as they lead to the same dual bounds, is based on Theorem 10.3, Proposition 11.1, and the fact that an exact separation algorithm for $\text{conv}(X^k)$ implicitly generates $\text{conv}(X^k)$.

Theorem 11.2 $z^{LPM} = w_{LD} = z^{CUT}$.

As the subproblems solved in both the column generation and Lagrangian dual approaches are optimization problems over X^k, column generation can be viewed as an algorithm for solving the Lagrangian dual in which the dual variables π are updated using linear programming by solving the Restricted Linear Programming Master. This is in comparison with the subgradient algorithm often used to solve the Lagrangian dual that is based on a much simpler updating procedure.

On the other hand, if we use the cutting plane approach, though the bound obtained is potentially the same, separation problems over $\text{conv}(X^k)$ have to be solved instead of optimization problems.

As the theoretical complexity of the optimization and separation problems for $\text{conv}(X^k)$ is the same, the choice of approach depends on the relative difficulty in solving the two problems as well as on the convergence of the column generation and cutting plane algorithms in practice.

11.4 IP COLUMN GENERATION FOR 0–1 IP

If when the column generation algorithm terminates, the optimal solution vector $\tilde{\lambda} = (\tilde{\lambda}^1, \ldots, \tilde{\lambda}^K)$ of LPM is not integer, then IPM is not yet solved. However, $z^{LPM} \geq z$, which suggests the possibility of using such upper bounds in a branch-and-bound algorithm. In this section we present an algorithm for 0–1 problems using this bound, called an *IP column generation* or *branch-and-price* algorithm.

Again we have the original problem

$$(IP) \quad \begin{array}{l} z = \max\{\sum_{k=1}^{K} c^k x^k : \sum_{k=1}^{K} A^k x^k = b, \\ D^k x^k \leq d^k \text{ for } k = 1, \ldots, K, \ x^k \in B^{n_k} \text{ for } k = 1, \ldots, K\}, \end{array}$$

and its reformulation

$$(IPM) \quad \begin{array}{l} z = \max \sum_{k=1}^{K} \sum_{t=1}^{T_k} (c^k x^{k,t}) \lambda_{k,t} \\ \sum_{k=1}^{K} \sum_{t=1}^{T_k} (A^k x^{k,t}) \lambda_{k,t} = b \\ \sum_{t=1}^{T_k} \lambda_{k,t} = 1 \text{ for } k = 1, \ldots, K \\ \lambda_{k,t} \in \{0, 1\} \text{ for } t = 1, \ldots, T_k, k = 1, \ldots, K. \end{array}$$

whose linear programming relaxation has optimal solution $\tilde{\lambda}$.

Because the points $x^{k,t} \in X^k$ are distinct 0–1 vectors, note that $\tilde{x}^k = \sum_{t=1}^{T_k} \tilde{\lambda}_{k,t} x^{k,t} \in \{0,1\}^{n_k}$ if and only if $\tilde{\lambda}$ is integer. Therefore if $\tilde{\lambda}$ is not integer, there is some κ and j such that the corresponding 0–1 variable x_j^κ has linear programming value $\tilde{x}_j^\kappa$ that is fractional, and on which one can branch.

This suggests the branching scheme shown in Figure 11.1(a), in which the set S of all feasible solutions is split into $S_0 = S \cap \{x : x_j^\kappa = 0\}$ and $S_1 = S \cap \{x : x_j^\kappa = 1\}$. Note that this is exactly the same type of scheme used in the basic branch-and-bound algorithm in Chapter 7.

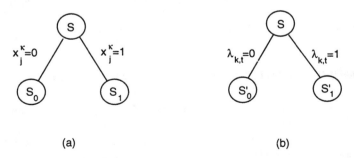

(a)   (b)

Fig. 11.1 Branching for 0–1 column generation: (a) original (b) column variables

It is important now to make sure that it will still be possible to solve the new linear programming master problems without difficulty. To do this we need to define the new Master Problems associated with S_i for $i = 0, 1$, and the new subproblems.

Now as $x_j^k = \sum_{t=1}^{T_k} \lambda_{k,t} x_j^{k,t} \in \{0,1\}$, $x_j^k = \delta \in \{0,1\}$ implies that $x_j^{k,t} = \delta$ for all k, t with $\lambda_{k,t} > 0$. So the Master Problem at node $S_i = S \cap \{x : x_j^\kappa = i\}$ for $i = 0, 1$ is

$$(IPM(S_i)) \quad \begin{aligned} z(S_i) = \max \sum_{k \neq \kappa} \sum_t (c^k x^{k,t}) \lambda_{k,t} + \sum_{t: x_j^{\kappa,t} = i} (c^\kappa x^{\kappa,t}) \lambda_{\kappa,t} \\ \sum_{k \neq \kappa} \sum_t (A^k x^{k,t}) \lambda_{k,t} + \sum_{t: x_j^{\kappa,t} = i} (A^\kappa x^{\kappa,t}) \lambda_{\kappa,t} = b \\ \sum_t \lambda_{k,t} = 1 \text{ for } k \neq \kappa \\ \sum_{t: x_j^{\kappa,t} = i} \lambda_{\kappa,t} = 1 \\ \lambda_{k,t} \in \{0,1\} \text{ for } t = 1, \ldots, T_k, k = 1, \ldots, K. \end{aligned}$$

This has the same form as the original Master Problem except that a set of columns are excluded on each branch, and the previous LPM solution is now infeasible. Turning to the column generation subproblems, the subproblem is unchanged if $k \neq \kappa$. However, for subproblem κ and $i = 0, 1$, we have

$$\zeta_\kappa(S_i) = \max\{(c^\kappa - \pi A^\kappa)x - \mu_\kappa : x \in X^\kappa, x_j = i\},$$

which is very similar to the original subproblem.

Another idea is to branch on some fractional $\lambda_{k,t}$ variable, fixing it to 0 and 1 respectively, see Figure 11.1(b). Note, however, that on the branch in which $\lambda_{k,t} = 0$, just one column, corresponding to the tth solution of subproblem k, is excluded, so the resulting problem is almost identical to the original one. This means that the resulting enumeration tree has the undesirable property of being highly unbalanced. In addition it is often difficult to impose the condition $\lambda_{k,t} = 0$, and thus to prevent the same solution being generated again as optimal solution after branching.

One potential advantage of the column generation approach, visible in Example 11.2, is that the optimal solutions to $RLPM$ are often integral or close to integral. In the first case this gives a feasible integer solution, and in the second such a solution can often be obtained by a simple rounding heuristic.

11.5 IMPLICIT PARTITIONING/PACKING PROBLEMS

An important subclass of decomposable 0–1 IPs are packing and partitioning problems. Given a finite set $M = \{1, \ldots, m\}$, there are K implicitly described sets of feasible subsets, and the problem is to find a maximum value packing or partition of M consisting of certain of these subsets.

In terms of the original IP (11.5) of Section 11.2, we set $x^k = (y^k, w^k)$ with $y^k \in \{0,1\}^m$ the incidence vector of subset k of M, $c^k = (e^k, f^k)$, $A^k = (I, 0)$ and $b = 1$. One should think of the variables w^k as auxiliary variables needed to define whether the subset with incidence vector y^k is feasible, and to define the possibly nonlinear objective value of the corresponding subset. So we have the formulation

$$z = \max\{\sum_{k=1}^{K}(e^k y^k + f^k w^k) : \sum_{k=1}^{K} y^k \leq 1, (y^k, w^k) \in X^k \text{ for } k = 1, \ldots, K\}.$$

Now if $(y^{k,t}, w^{k,t})$ corresponds to the tth feasible solution in the set X^k, and $\lambda_{k,t}$ is the corresponding variable, we obtain an equivalent Integer Programming Master

$$z = \max \sum_{k=1}^{K} \sum_{t=1}^{T_k} (e^k y^{k,t} + f^k w^{k,t}) \lambda_{k,t}$$
$$\sum_{k=1}^{K} \sum_{t: y_i^{k,t}=1} \lambda_{k,t} = 1 \text{ for } i \in M$$
$$\sum_{t=1}^{T_k} \lambda_{k,t} \leq 1 \text{ for } k = 1, \ldots, K$$
$$\lambda_{k,t} \in \{0,1\} \text{ for } t = 1, \ldots, T_k, k = 1, \ldots, K.$$

We now present several problems of this type. Clearly as the partitioning problem is a special case of (11.5), the algorithm of the previous section can be applied.

Multi-Item Lot-Sizing. Suppose we are given demands d_t^k for items $k = 1, \ldots, K$ over a time horizon $t = 1, \ldots, T$. All items must be produced on a single machine; the machine can produce only one item in each period and has a capacity C_t^k if item k is produced in period t. Given production, storage, and set-up costs for each item in each period, we wish to find a minimum cost production plan. This problem can be formulated as

$$\min \sum_{k=1}^{K} \sum_{t=1}^{T} (p_t^k x_t^k + h_t^k s_t^k + f_t^k y_t^k)$$
$$\sum_{k=1}^{K} y_t^k \leq 1 \text{ for } t = 1, \ldots, n$$
$$(x^k, s^k, y^k) \in X^k \text{ for } k = 1, \ldots, K$$

where $X^k = \{(x^k, s^k, y^k) \in R_+^n \times R_+^n \times B^n : s_{t-1}^k + x_t^k = d_t^k + s_t^k, x_t^k \leq C_t^k y_t^k \text{ for } t = 1, \ldots, n\}$.

Clustering. Given a graph $G = (V, E)$, edge costs c_e for $e \in E$, node weights d_i for $i \in V$, and a cluster capacity C, we wish to split the node set V into K (possibly empty) clusters satisfying the property that the sum of the node weights in each cluster does not exceed C, in a way that minimizes the sum of the weights of edges between clusters (maximizes the sum of weights of edges within clusters). Figure 11.2 shows a feasible solution for an instance with 3 clusters and a capacity of 9. The thick edges are those between clusters. The problem can be formulated as

$$\max \sum_{k=1}^{K} \sum_{e \in E} c_e w_e^k$$
$$\sum_{k=1}^{K} y_i^k \leq 1 \text{ for } i \in V$$
$$(w^k, y^k) \in X^k \text{ for } k = 1, \ldots, K$$

where $X^k = \{(w^k, y^k) \in B^m \times B^n : w_e^k \leq y_i^k, w_e^k \leq y_j^k, w_e^k \geq y_i^k + y_j^k - 1$ for $e = (i,j) \in E, \sum_{i \in V} d_i y_i^k \leq C\}$ with $y_i^k = 1$ if node i is in cluster k and $w_e^k = 1$ if edge e has both endpoints in cluster k.

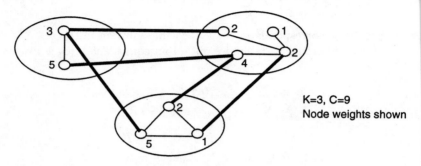

K=3, C=9
Node weights shown

Fig. 11.2 Clustering solution with three clusters

Capacitated Vehicle Routing. Given a graph $G = (V, E)$, a depot node 0, edge costs c_e for each $e \in E$, K identical vehicles of capacity C, and client orders d_i for $i \in V \setminus \{0\}$, we wish to find a set of subtours (cycles) for each vehicle such that (i) each subtour contains the depot, (ii) together the subtours contain all the nodes, (iii) the subtours are disjoint on the node set $V \setminus \{0\}$, and (iv) the total demand on each subtour (the total amount delivered by each vehicle) does not exceed C.

Another problem with such a decomposable structure is the generalized assignment problem. An instance of *GAP* is treated by branch-and-cut in Section 9.6.

11.6 PARTITIONING WITH IDENTICAL SUBSETS*

The clustering and vehicle routing problems of the last section both have the property that the clusters or vehicles are interchangeable (independent of k). This means that the numbering of the subsets is arbitrary, and exchanging any two sets leads to an essentially identical solution.

Here we consider how the integer programming column generation algorithm of Section 11.4 can be specialized to take account of this symmetry. As $X^k = X$, $(e^k, f^k) = (e, f)$ and $T_k = T$ for all k, we can set $\lambda_t = \sum_{k=1}^{K} \lambda_{k,t}$ and *IPM* now takes the form:

$$\max \sum_{t=1}^{T} (ey^t + fw^t) \lambda_t$$
$$\sum_{t: y_i^t = 1} \lambda_t = 1 \text{ for } i \in M$$

$$\sum_{t=1}^{T} \lambda_t \leq K$$
$$\lambda \in B^T.$$

There is now just a single column generation subproblem. Letting the dual variables associated with the linear programming relaxation be $\{\pi_i\}_{i \in M}$ and μ, the subproblem is:

$$\zeta = \max\{(e - \pi)y + fw - \mu : (y, w) \in X\}$$

and LPM can be solved as in Section 11.3.

What happens if the solution $\tilde{\lambda}$ of LPM is not integral? It is now not at all obvious how to recover the original variables x^k or the $\lambda_{k,t}$ variables, so the branching scheme proposed in Section 11.4 must be modified. We now consider two possibilities.

Branching Rules

(i) If $\sum_{t=1}^{T} \lambda_t = \alpha \notin Z$, then form two branches with $\sum_{t=1}^{T} \lambda_t \leq \lfloor \alpha \rfloor$ and $\sum_{t=1}^{T} \lambda_t \geq \lceil \alpha \rceil$ respectively.

(ii) A second possibility is based on the simple observation that if we take two elements of M, either they appear together in some subset, or not. So we choose a pair of elements (rows) i and j in M for which

$$0 < \sum_{t: y_i^t = y_j^t = 1} \lambda_t < 1,$$

and we then form two branches with $\sum_{t: y_i^t = y_j^t = 1} \lambda_t = 1$ and $\sum_{t: y_i^t = y_j^t = 1} \lambda_t = 0$ respectively.

In the first case (i) we impose that i and j lie in the same subset, and in the second case (ii) that they lie in different subsets. In case (i) all columns corresponding to subsets Q containing either i or j but not both are eliminated from the Master Problem, and the constraint $y_i = y_j$ is added to the subproblem to ensure that any new column generated does not generate a subset containing i but not j, or vice versa. In case (ii) columns containing both i and j are eliminated from the Master, and the constraint $y_i + y_j \leq 1$ is added to the subproblem.

So, imposing the constraints $y_i = y_j$ or $y_i + y_j \leq 1$ on each subproblem permits us to branch as shown in Figure 11.3.

The following result says that this second branching scheme is sufficient.

Proposition 11.3 *If $\tilde{\lambda} \notin B^T$, there exist rows $i, j \in M$ such that*

$$0 < \sum_{t: y_i^t = y_j^t = 1} \lambda_t < 1.$$

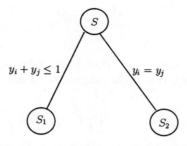

Fig. 11.3 A branching scheme for partitioning

Such a pair is also not difficult to find.

Example 11.2 Consider an instance of the clustering problem of Section 11.5 with $G = (V, E)$ the complete graph on 3 nodes, $K = 3$ clusters, the objective of choosing as many edges as possible within the clusters, and at most 2 nodes allowed per cluster, that is, node weights $d_i = 1$ for all $i \in V$, edge weights $c_e = 1$ for all $e \in E$, and cluster capacity $C = 2$.

1. *Solving LPM.* Starting the Restricted LPM with clusters consisting of single nodes leads to

$$\begin{aligned}
\max 0\lambda_1 + 0\lambda_2 + 0\lambda_3 & \\
1\lambda_1 + 0\lambda_2 + 0\lambda_3 &= 1 \\
0\lambda_1 + 1\lambda_2 + 0\lambda_3 &= 1 \\
0\lambda_1 + 0\lambda_2 + 1\lambda_3 &= 1 \\
1\lambda_1 + 1\lambda_2 + 1\lambda_3 &\leq 3 \\
\lambda &\geq 0
\end{aligned}$$

with $RLPM$ value 0, primal solution $\lambda = (1,1,1)$, and dual solution $\pi = (0,0,0), \mu = 0$. This provides a feasible solution to the original problem of objective value 0, and so we set $\underline{z} = 0$.

2. *Solving the Subproblem.* The subproblem of selecting a feasible cluster of maximum reduced price is

$$\begin{aligned}
\zeta = \min w_{12} + w_{13} + w_{23} - 0y_1 - 0y_2 - 0y_3 - 0 & \\
w_{12} \leq y_1, w_{12} \leq y_2, w_{12} &\geq y_1 + y_2 - 1 \\
w_{13} \leq y_1, w_{13} \leq y_3, w_{13} &\geq y_1 + y_3 - 1 \\
w_{23} \leq y_2, w_{23} \leq y_3, w_{23} &\geq y_2 + y_3 - 1 \\
y_1 + y_2 + y_3 &\leq 2 \\
w \in B^{|E|}, y &\in B^{|V|}
\end{aligned}$$

giving $\zeta = 1$ and an optimal solution $w_{12} = y_1 = y_2 = 1$.

3. *Solution of LPM.* After three iterations LPM is solved in the form

$$\max 0\lambda_1 + 0\lambda_2 + 0\lambda_3 + 1\lambda_4 + 1\lambda_5 + 1\lambda_6$$
$$1\lambda_1 + 0\lambda_2 + 0\lambda_3 + 1\lambda_4 + 1\lambda_5 + 0\lambda_6 = 1$$
$$0\lambda_1 + 1\lambda_2 + 0\lambda_3 + 1\lambda_4 + 0\lambda_5 + 1\lambda_6 = 1$$
$$0\lambda_1 + 0\lambda_2 + 1\lambda_3 + 0\lambda_4 + 1\lambda_5 + 1\lambda_6 = 1$$
$$1\lambda_1 + 1\lambda_2 + 1\lambda_3 + 1\lambda_4 + 1\lambda_5 + 1\lambda_6 \leq 3$$
$$\lambda \geq 0$$

with optimal solution $\lambda_4 = \lambda_5 = \lambda_6 = \frac{1}{2}$ and $z^{LPM} = \frac{3}{2}$.

4. *Branching.* Taking rows $i = 1$ and $l = 2$, we use the second branching scheme and split the problem into two subproblems:

S_1 is the set of solutions in which nodes 1 and 2 do not lie in the same cluster, so S_1 is obtained by setting $\lambda_4 = 0$ cutting off the existing solution. All new clusters containing both nodes 1 and 2 are excluded.

S_2 is the set of solutions in which any cluster containing either 1 or 2 must contain the other, so S_2 is obtained by setting $\lambda_2 = \lambda_3 = \lambda_5 = \lambda_6 = 0$ cutting off the existing solution. Any new clusters containing just one of the nodes 1,2 are excluded.

5. *Reoptimizing for S_1.* With $\lambda_4 = 0$, the new $RLPM$ is

$$\max 0\lambda_1 + 0\lambda_2 + 0\lambda_3 + 1\lambda_5 + 1\lambda_6$$
$$1\lambda_1 + 0\lambda_2 + 0\lambda_3 + 1\lambda_5 + 0\lambda_6 = 1$$
$$0\lambda_1 + 1\lambda_2 + 0\lambda_3 + 0\lambda_5 + 1\lambda_6 = 1$$
$$0\lambda_1 + 0\lambda_2 + 1\lambda_3 + 1\lambda_5 + 1\lambda_6 = 1$$
$$1\lambda_1 + 1\lambda_2 + 1\lambda_3 + 1\lambda_5 + 1\lambda_6 \leq 3$$
$$\lambda \geq 0$$

with optimal primal solution $\lambda_1 = \lambda_6 = 1$, and dual solution $\pi = (0, 0, 1), \mu = 0$. The incumbent is updated, $\underline{z} = 1$.

6. *Subproblem for S_1.* The subproblem is

$$\zeta = \min w_{12} + w_{13} + w_{23} - 1y_1 - 0y_2 - 0y_3 - 0$$
$$w_{12} \leq y_1, w_{12} \leq y_2, w_{12} \geq y_1 + y_2 - 1$$
$$w_{13} \leq y_1, w_{13} \leq y_3, w_{13} \geq y_1 + y_3 - 1$$
$$w_{23} \leq y_2, w_{23} \leq y_3, w_{23} \geq y_2 + y_3 - 1$$
$$y_1 + y_2 + y_3 \leq 2$$
$$y_1 + y_2 \leq 1$$
$$w \in B^{|E|}, y \in B^{|V|}$$

giving $\zeta = 0$. So $LPM(S_1)$ is solved with $z^{LPM}(S_1) = 1$. The node is pruned by bound.

7. *Node S_2*. When setting $\lambda_2 = \lambda_3 = \lambda_5 = \lambda_6 = 0$, the new $RLPM$ has unique optimal solution $\lambda_3 = \lambda_4 = 1$. We continue iterating between subproblem and the restricted Master till LPM is solved with $z^{LPM}(S_2) = 1$. Then the node is pruned by bound, and as there are no outstanding nodes, the incumbent solution $\lambda_1 = \lambda_6 = 1$ is optimal. This corresponds to one cluster containing node 1, another containing nodes 2,3, and the third necessarily empty. ∎

11.7 NOTES

11.1 The fundamental paper on the decomposition of linear programs, known as Dantzig-Wolfe decomposition, is [DanWol60]. Recent surveys in this area include [Barnetal94] and [Desetal95].

11.3 The first use of column generation to solve the Master linear program arising from an integer programming problem is probably the work on the cutting stock problem [GilGom61], [GilGom63]. The equivalence of the bounds provided by the linear programming Master and the Lagrangian dual has been known since [Geo74].

11.4 The first papers on integer programmingt column generation appeared in the eighties [DesSouDes84],[DesSou89] on routing problems in which the subproblems are constrained shortest path problems that are solved by dynamic programming.

11.5 The multi-item lot-sizing and clustering problems have been tackled by integer programming decomposition in [Vdbeck94], and the clustering problem in [JohMehNem93], the generalized assignment problem in [Sav93], and binary and integer cutting stock problems in [Vancetal94] and [Vdbeck96] respectively.

11.6 The branching rule (ii) is from [RyaFos81]. Recent more general branching rules that are not restricted to 0–1 problems appear in [Barnetal94] and [VdbeckWol96].

In [Ben62] an alternative resource-based reformulation and decomposition approach is proposed; see Exercise 11.5.

11.8 EXERCISES

1. Consider the following instance of UFL with $m = 4, n = 3$,

$$(c_{ij}) = \begin{pmatrix} 2 & 1 & 5 \\ 3 & 4 & 2 \\ 6 & 4 & 1 \\ 1 & 3 & 7 \end{pmatrix} \text{ and } f = (8, 6, 5).$$

Reformulate using an Integer Programming Master Problem. Solve the Linear Programming Master by column generation.

2. Solve the following instance of $STSP$ by column generation

$$(c_e) = \begin{pmatrix} - & 3 & 4 & 2 \\ - & - & 5 & 6 \\ - & - & - & 12 \\ - & - & - & - \end{pmatrix}.$$

3. Consider GAP with equality constraints

$$\max \sum_{i=1}^m \sum_{j=1}^n c_{ij} x_{ij}$$
$$\sum_{j=1}^n x_{ij} = 1 \text{ for } i = 1, \ldots, m$$
$$\sum_{i=1}^m a_{ij} x_{ij} \leq b_j \text{ for } j = 1, \ldots, n$$
$$x \in B^{mn}.$$

Solve an instance with $m = 3, n = 2$, $(c_{ij}) = (a_{ij}) = \begin{pmatrix} 2 & 1 \\ 1 & 1 \\ 1 & 2 \end{pmatrix}$, and $b = \begin{pmatrix} 2 \\ 2 \end{pmatrix}$ by inter programming decomposition. Solve also by Lagrangian relaxation and by cutting planes and compare.

4. Formulate the Integer Programming Master and subproblems for the three problems presented in Section 11.5.

5.* (Benders' Reformulation). Use the results of Exercise 1.15 to show that

(MIP) $z = \max\{cx + hy : Ax + Gy \leq b, x \in R_+^n, y \in Y \subseteq R_+^p\}$

has the equivalent formulation

$$z = \max \eta$$
$$\eta \leq u^s(b - Gy) + hy \text{ for } s = 1, \ldots, S$$
$$v^t(b - Gy) \geq 0 \text{ for } t = 1, \ldots, T$$
$$y \in Y.$$

Describe a cutting plane algorithm for MIP based on this reformulation.

6. Consider the problem of scheduling n jobs on m identical machines. The processing time of job j is p_j for $j = 1, \ldots, n$, and the objective is to terminate all jobs as soon as possible. Formulate as an IP and discuss algorithmic options for the problem.

7. Suppose that 15 pieces of length 32, 35 of length 20, 17 of length 15, and 42 of length 11 must be cut from sheets of length 104. Find the minimum number of sheets required. Formulate, discuss possible algorithms, and present lower and upper bounds on the minimum value.

12
Heuristic Algorithms

12.1 INTRODUCTION

Given that many, if not most, of the practical problems that we wish to solve are $\mathcal{NP}$-hard, it is not surprising that heuristic or approximation algorithms play an important role in "solving" discrete optimization problems—the idea being to hopefully find a "good" feasible solution quickly.

Different reasons may lead one to choose a heuristic:

A solution is required rapidly, within a few seconds or minutes.

The instance is so large and/or complicated that it cannot be formulated as an *IP* or *MIP* of reasonable size.

Even though it has been formulated as an *MIP*, it is difficult or impossible for the branch-and-bound system to find (good) feasible solutions.

For certain combinatorial problems such as vehicle routing and machine scheduling, it is easy to find feasible solutions by inspection or knowledge of the problem structure, and a general-purpose mixed integer programming approach is ineffective.

In designing and using a heuristic, there are various questions one can ask:

Should one just accept any feasible solution, or should one ask *a posteriori* how far it is from optimal?

Can one guarantee *a priori* that the heuristic will produce a solution within ϵ (or $\alpha\%$) of optimal?

204   HEURISTIC ALGORITHMS

Can one say *a priori* that, for the class of problems considered, the heuristic will on average produce a solution within $\alpha\%$ of optimal?

The rest of this chapter is divided into four parts. In the first, we formalize the greedy and local exchange heuristics introduced by example in Chapter 2. We then consider two improved local exchange heuristics, tabu search and simulated annealing, that include ways to escape from a local optimum, and genetic algorithms that work with families of solutions. These heuristics, though often very effective, provide no (dual) performance bounds and thus no direct way of assessing the quality of the solutions found. We then consider some problems for which by simple analysis a priori worst-case bounds can be obtained. Though these bounds are typically weak, only guaranteeing solutions within, say 50%, of optimal, the resulting heuristics are typically much more effective in practice. Finally we discuss how to use a mixed integer programming system in heuristic fashion with the aim of finding good solutions quickly, and also of obtaining at least some *a posteriori* performance guarantees.

12.2 GREEDY AND LOCAL SEARCH REVISITED

Here we formalize the greedy and local search algorithms presented by example in Section 2.6. First we suppose that the problem can be written as a combinatorial problem in the form:

$$\min_{S \subseteq N} \{c(S) : v(S) \geq k\}.$$

For example, the 0–1 knapsack problem

$$\min\{\sum_{j=1}^{n} c_j x_j : \sum_{j=1}^{n} a_j x_j \geq b, \ x \in B^n\}$$

with $c_j, a_j \geq 0$ for $j = 1,\ldots,n$ is of this form with $c(S) = \sum_{j \in S} c_j$, $v(S) = \sum_{j \in S} a_j$ and $k = b$. The uncapacitated facility location problem also fits this model if we take $c(S) = \sum_{i \in M} \min_{j \in S} c_{ij} + \sum_{j \in S} f_j$ for $S \neq \emptyset$, $v(S) = |S|$ and $k = 1$.

Below we assume that the empty set is infeasible.

A Greedy Heuristic

1. Set $S^0 = \emptyset$ (start with the empty set). Set $t = 1$.
2. Set $j_t = \arg\min \frac{c(S^{t-1} \cup \{j_t\}) - c(S^{t-1})}{v(S^{t-1} \cup \{j_t\}) - v(S^{t-1})}$ (choose the element whose additional cost per unit of resource is minimum).
3. If the previous solution S^{t-1} is feasible, and the cost has not decreased, stop with $S^G = S^{t-1}$.

4. Otherwise set $S^t = S^{t-1} \cup \{j_t\}$. If the solution is now feasible, and the cost function is nondecreasing or $t = n$, stop with $S^G = S^t$.
5. Otherwise if $t = n$, no feasible solution has been found. Stop.
6. Otherwise set $t \leftarrow t + 1$, and return to 2.

Example 12.1 We apply the greedy heuristic to an instance of the uncapacitated facility location problem with $m = 6$ clients, $n = 4$ depots, and costs

$$(c_{ij}) = \begin{pmatrix} 6 & 2 & 3 & 4 \\ 1 & 9 & 4 & 11 \\ 15 & 2 & 6 & 3 \\ 9 & 11 & 4 & 8 \\ 7 & 23 & 2 & 9 \\ 4 & 3 & 1 & 5 \end{pmatrix} \text{ and } f = (21, 16, 11, 24).$$

The algorithm gives:

Initialization. $S^0 = \phi$. S^0 is infeasible.

Iteration 1. $c(1) = (6 + 1 + 15 + 9 + 7 + 4) + 21 = 63, c(2) = 66, c(3) = 31, c(4) = 64$, so $j_1 = 3$, $S^1 = \{3\}$ and $c(S^1) = 31$. S^1 is feasible.

Iteration 2. $c(1,3) - c(3) = (\min\{6,3\} + \min\{1,4\} + \min\{15,6\} + \min\{9,4\} + \min\{7,2\} + \min\{4,1\}) + (21+11) - 31 = 18, c(2,3) - c(3) = 11, c(3,4) - c(3) = 11$.

Termination. As the cost has not decreased, the algorithm stops with heuristic solution $S^G = \{3\}$ and cost 31. ∎

Greedy heuristics have to be adapted to the particular problem structure. For $STSP$ there are several possible greedy heuristics that choose edges one after another until a tour is obtained. The "nearest neighbor" heuristic starts from some arbitrary node, and then greedily constructs a path out from that node. The "pure greedy" heuristic chooses a least-cost edge j_t such that S^t is still part of a tour (i.e., S^t consists of a set of disjoint paths, until the last edge chosen forms a tour).

To describe local search it is simpler to formulate the combinatorial optimization problem as

$$\min_{S \subseteq N} \{c(S) : g(S) = 0\}$$

where $g(S) \geq 0$ represents a measure of the infeasibility of set S. Thus the constraint $v(S) \geq k$ used above can be represented here by $g(S) = (k - v(S))^+$.

For a local search algorithm, we need to define a *solution*, a *local neighborhood* $Q(S)$ for each solution $S \subseteq N$, and a *goal function* $f(S)$ which can either be just equal to $c(S)$ when S is feasible, and infinite otherwise, or a composite function of the form $c(S) + \alpha g(S)$ consisting of a weighted combination of the objective function value and a positive multiple α of the infeasibility measure for S.

A Local Search Heuristic. Choose an initial solution S. Search for a set $S' \in Q(S)$ with $f(S') < f(S)$. If none exists, stop. $S^H = S$ is a local optimum solution.

Otherwise set $S = S'$, and repeat.

Appropriate choices of neighborhood depend on the problem structure. A very simple neighborhood is that in which just one element is added or removed from S, that is, $Q(S) = \{S' : S' = S \cup \{j\} \text{ for } j \in N \setminus S\} \cup \{S' : S' = S \setminus \{i\} \text{ for } i \in S\}$. This neighborhood has only $O(n)$ elements.

Another neighborhood that is appropriate if feasible sets all have the same size is that in which one element of S is replaced by another element not in S, that is, $Q(S) = \{S' : S' = S \cup \{j\} \setminus \{i\} \text{ for } j \in N \setminus S \text{ and } i \in S\}$. This neighborhood has $O(n^2)$ elements.

For $STSP$, there is no tour differing from an initial tour by a single edge. However, if two edges are removed, there is exactly one other tour containing the remaining edges (see Figure 12.1). This leads to the well-known 2-exchange heuristic for the $STSP$ on a complete graph.

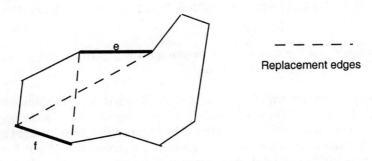

Fig. 12.1 2-Exchange for STSP

2-Exchange Heuristic for STSP

The local search heuristic is applied with the following specifications:
A set $S \subset E$ is a solution if the set of edges S form a tour.
$Q(S) = \{S' \text{ is a solution: } S' \neq S, |S' \cap S| = n - 2\}$, where $n = |V|$.
$f(S) = \sum_{e \in S} c_e$.
The resulting local search solution is called a *2-optimal tour*.

12.3 IMPROVED LOCAL SEARCH HEURISTICS

How do we escape from a local minimum, and thus potentially do better than a local search heuristic? This is the question addressed by the tabu and simulated annealing heuristics that we now present briefly.

12.3.1 Tabu Search

When at a local minimum, a natural idea is to move to the best solution in the neighborhood even though its value is worse. One obvious difficulty with this idea is that cycling may occur, that is, the algorithm returns to the same solution every two or three steps: $S^0 \to S^1 \to S^0 \to S^1 \ldots$.

To avoid such cycling, certain solutions or moves are *forbidden* or *tabu*. Directly comparing the new solution with a list of all previous incumbents would require much space and be very time consuming. Instead, a *tabu list* of recent solutions, or recent solution modifications, is kept.

A basic version of the tabu search algorithm can be described as follows:

1. Initialize an empty tabu list.
2. Get an initial solution S.
3. While the stopping criterion is not satisfied:
   3.1. Choose a subset $Q'(S) \subseteq Q(S)$ of non-tabu solutions.
   3.2. Let $S' = \arg\min\{f(T) : T \in Q'(S)\}$.
   3.3. Replace S by S' and update the tabu list.
4. On termination, the best solution found is the heuristic solution.

The parameters specific to tabu search are:
(i) The choice of subset $Q'(S)$. Here if $Q(S)$ is small, one takes the whole neighborhood, while if $Q(S)$ is large, $Q'(S)$ can be a fixed number of neighbors of S, chosen randomly or by some heuristic rule.

(ii) The tabu list consists of a small number t of most recent solutions or modifications. If $t = 1$ or 2, it is not surprising that cycling is still common. The magic value $t = 7$ is often cited as a good choice.

(iii) The stopping rule is often just a fixed number of iterations, or a certain number of iterations without any improvement of the goal value of the best solution found.

Considering the neighborhood function

$$Q(S) = \{T \subseteq V : T = S \cup \{j\} \text{ for } j \in V \setminus S\} \cup \{T \subseteq V : T = S \setminus \{i\} \text{for } i \in S\}$$

consisting of single element switches, the tabu list might be a list of the last t elements $\{i_1, \ldots, i_t\}$ to be added to an incumbent, and of the last t elements $\{j_1, \ldots, j_t\}$ to be removed. A neighbor T is then **tabu** if $T = S \setminus \{i_q\}$ for some $q = 1,\ldots,t$ or if $T = S \cup \{j_q\}$ for some $q = 1,\ldots,t$. Therefore one cannot remove one of the t elements added most recently, and one cannot add one of the t elements removed most recently.

Tabu search also uses common sense. There is no justification to make a solution tabu if it is the best solution found to date, or it is interesting for some reason. So one or more *aspiration levels* can be defined that are used to overrule the tabu criteria. More generally, tabu search can be viewed as a search strategy that tries to take advantage of the history of the search and the problem structure intelligently.

12.3.2 Simulated Annealing

Simulated annealing is less direct. The basic idea is to choose a neighbor randomly. The neighbor then replaces the incumbent with probablity 1 if it has a better goal value, and with some probability strictly between 0 and 1 if it has a worse goal value.

The probability of accepting a worse solution is proportional to the difference in goal values, so slightly worse solutions have a high probability of being accepted, while much worse solutions will only be accepted infrequently. Therefore if the number of iterations is sufficiently large, it means that one can move away from any local minimum. On the other hand, for the process to converge in the long run, the probability of accepting worse solutions decreases over time, so the algorithm should end up converging to a "good" local minimum.

A Simulated Annealing Heuristic

1. Get an initial solution S.
2. Get an initial temperature T and a reduction factor r with $0 < r < 1$.
3. While not yet frozen, do the following:
    3.1 Perform the following loop L times:
        3.1.1 Pick a random neighbor S' of S.
        3.1.2 Let $\Delta = f(S') - f(S)$.
        3.1.3 If $\Delta \leq 0$, set $S = S'$.
        3.1.4 If $\Delta > 0$, set $S = S'$ with probability $e^{-\Delta/T}$.
    3.2 Set $T \leftarrow rT$. (Reduce the temperature.)
4. Return the best solution found.

Note that as specified above, the larger Δ is, the less chance there is of making a move to a solution worse by Δ. Also as the temperature decreases, the chances of making a move to a worse solution decrease.

Exactly as for local exchange heuristics, one has to define:

(i) A solution

(ii) The neighbors of a solution
(iii) The cost of a solution
(iv) How to determine an initial solution.

The other parameters specific to simulated annealing are then:

(v) The initial temperature T
(vi) The cooling ratio r
(vii) The loop length L
(viii) The definition of "frozen," or the stopping criterion.

As application, we again consider the graph equipartition problem (see Section 2.6). There we defined a solution S to be an equipartition $(S, V \setminus S)$ with the two sets differing in size by at most one, and a neighborhood $Q(S) = \{T \subset V : |T \setminus S| = |S \setminus T| = 1\}$.

Here we go for more flexibility, by allowing any set $S \subseteq V$ representing the partition $(S, V \setminus S)$ to be a solution.

The neighborhood of a solution S is defined by a single element switch with $Q(S)$ as in Section 12.3.1. The cost of a partition is

$$f(S) = \sum_{e \in \delta(S, V \setminus S)} c_e + \alpha(|S| - |V \setminus S|)^2$$

for some $\alpha > 0$. Therefore any disparity in the size of the two sets is penalized in the goal function.

In designing and discussing improved local search algorithms, three more general concepts are useful in thinking about the right combination of choices.

Communication. It is important that the neighborhood structure be such that it is possible to get from any solution S to any other solution S' preferably in a small number of moves. Failing this, it should be possible to get from any solution S to at least one optimal solution.

Diversification. This relates to facilitating movement between very different areas of the search space. A high initial temperature T, a long tabu list, and the possibility of using random restarts all encourage diversification.

Intensification. This relates to the opposite idea of increasing the search effort in promising areas of the search space. Choosing optimally in the neighborhood, or enlarging the set $Q'(S)$ of neighbors temporarily, are measures of intensification.

12.3.3 Genetic Algorithms

Rather than working to improve individual solutions, genetic algorithms work with a finite *population* (set of solutions) $S_1, \ldots, S_k$, and the population evolves (changes somewhat randomly) from one *generation* (iteration) to the next.

An iteration consists of the following steps:

(i) **Evaluation.** The *fitness* of the individuals is evaluated.

(ii) **Parent Selection.** Certain pairs of solutions (*parents*) are selected based on their fitness.

(iii) **Crossover.** Each pair of parents combines to produce one or two new solutions (*offspring*).

(iv) **Mutation.** Some of the offspring are randomly modified.

(v) **Population Selection.** Based on their fitness, a new population is selected replacing some or all of the original population by an identical number of offspring.

We now indicate briefly ways in which the different steps can be carried out.

Evaluation. As in the local search algorithms, a goal function $f(S)$ or a pair of objective and infeasibility functions $c(S)$ and $g(S)$ are used to measure the fitness of a solution S.

Parent Selection. The idea is to choose "fitter" solutions with a higher probability. Thus from the initial population, S_i is chosen with probability $\frac{f(S_i)}{\sum_{j=1}^{k} f(S_j)}$.

The implementation of crossover and mutation are more problem dependent.

Crossover. Here one seeks some natural way to combine two fit solutions to produce a new fit solution. One way this is often done is by representing the solution S as a binary or integer string $x_1 x_2 \ldots x_r$.

Three possible ways to combine such strings are:

1-point crossover. Given two strings $x_1 x_2 \ldots x_r$ and $y_1 y_2 \ldots y_r$, and an integer $p \in \{1, \ldots, r-1\}$, the two children are $x_1 \ldots x_p y_{p+1} \ldots y_r$ and $y_1 \ldots y_p x_{p+1} \ldots x_r$.

2-point crossover. Given two strings $x_1 x_2 \ldots x_r$ and $y_1 y_2 \ldots y_r$, and integers $p, q \in \{1, \ldots, r-1\}$ with $p < q$, the two children are $x_1 \ldots x_p y_{p+1} \ldots y_q x_{q+1} \ldots x_r$ and $y_1 \ldots y_p x_{p+1} \ldots x_q y_{q+1} \ldots y_r$.

Uniform crossover. Given two strings $x_1 x_2 \ldots x_r$ and $y_1 y_2 \ldots y_r$, the result is a child $z_1 \ldots z_r$ where each z_i is randomly chosen from $\{x_i, y_i\}$ for $i = 1, \ldots, r$.

Mutation. A simple 1-point mutation of a child $z_1 \ldots z_r$ is a random choice of $p \in \{1, \ldots, r\}$ and a random integer $\tilde{z}_p$ from the appropriate range giving a modified solution $z_1 \ldots z_{p-1} \tilde{z}_p z_{p+1} \ldots z_r$. A 2-point mutation is a swap of the values z_p and z_q for some $p < q$.

Consider again the generalized assignment problem in the form:

$$\min \sum_{i \in M} \sum_{j \in N} c_{ij} x_{ij} \qquad (12.1)$$
$$\sum_{j \in N} x_{ij} = 1 \text{ for } i \in M \qquad (12.2)$$
$$\sum_{i \in M} a_{ij} x_{ij} \leq b_i \text{ for } j \in N \qquad (12.3)$$
$$x \in B^{|M| \times |N|}. \qquad (12.4)$$

S is a "solution" if its incidence vector satisfies (12.2) and (12.4). It is represented by an integer m-vector $(j_1, \ldots, j_m)$ where j_i is the job to which person i is assigned (i.e, $x_{i,j_i} = 1$ for $i \in M$).

Two fitness values are the objective value $c(S) = \sum_{i \in M} c_{i,j_i}$ and the infeasibility value $g(S) = \sum_{j \in N} (\sum_{i \in M} a_{i,j_i} - b_j)^+$.

The two-point crossover and the two-point mutation described above, both lead to new solutions satisfying (12.2) and (12.4).

Finally, the objective value $c(S)$ might be used in the selection of parents, while the feasibility measure $g(S)$ may be appropriate for the selection of the new population.

These simple ideas are obviously far from covering all possibilities. For problems in which the solution is a permutation, such as *STSP* or machine scheduling, some more suitable form of crossover is necessary.

12.4 WORST-CASE ANALYSIS OF HEURISTICS

To start with, we consider a very simple example, the integer knapsack problem:

$$(IKP) \qquad z = \max\{\textstyle\sum_{j=1}^n c_j x_j : \sum_{j=1}^n a_j x_j \leq b, x \in Z_+^n\}$$

where $\{a_j\}_{j=1}^n, b \in Z_+$. Without loss of generality we assume that $a_j \leq b$ for $j \in N$ and $\frac{c_1}{a_1} \geq \frac{c_j}{a_j}$ for $j \in N \setminus \{1\}$.

The greedy heuristic solution for IKP is $x^H = (\lfloor \frac{b}{a_1} \rfloor, 0, \ldots, 0)$ with value $z^H = cx^H$.

212 HEURISTIC ALGORITHMS

Theorem 12.1 $\frac{z^H}{z} \geq \frac{1}{2}$.

Proof. The solution to the linear programming relaxation of IKP is $x_1 = \frac{b}{a_1}, x_j = 0$ for $j = 2, \ldots, n$ giving an upper bound $z^{LP} = \frac{c_1 b}{a_1} \geq z$.

Now as $a_1 \leq b$, $\lfloor \frac{b}{a_1} \rfloor \geq 1$. Setting $\frac{b}{a_1} = \lfloor \frac{b}{a_1} \rfloor + f$ with $0 \leq f < 1$, we have that $\lfloor \frac{b}{a_1} \rfloor / \frac{b}{a_1} = \frac{\lfloor \frac{b}{a_1} \rfloor}{\lfloor \frac{b}{a_1} \rfloor + f} \geq \frac{\lfloor \frac{b}{a_1} \rfloor}{\lfloor \frac{b}{a_1} \rfloor + \lfloor \frac{b}{a_1} \rfloor} = \frac{1}{2}$.

So $\frac{z^H}{z} \geq \frac{z^H}{z^{LP}} = \frac{c_1 \lfloor \frac{b}{a_1} \rfloor}{c_1 \frac{b}{a_1}} = \frac{\lfloor \frac{b}{a_1} \rfloor}{\frac{b}{a_1}} \geq \frac{1}{2}$. ∎

It is important to observe that the analysis depends on finding both a *lower bound* on z from the heuristic solution, and also an *upper bound* on z coming from a relaxation or dual solution.

As a second problem we consider $STSP$ on a complete graph with the edge lengths satisfying the *triangle inequality*, that is, if $e_i \in E$ for $i = 1, 2, 3$ are the three sides of a triangle, then $c_{e_i} + c_{e_j} \geq c_{e_k}$ for $i \neq j \neq k, i, j, k \in \{1, 2, 3\}$. Note that when this inequality holds, the shortest path between two nodes i, j is along the edge (i, j) (see Figure 12.2).

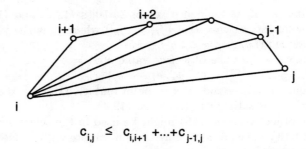

Fig. 12.2 Triangle inequality

To understand the heuristics and their analysis we need to introduce Eulerian graphs.

Definition 12.1 $G = (V, E)$ is a *Eulerian graph* if the degree of each node is even.

Proposition 12.2 *If $G = (V, E)$ is a connected Eulerian graph and $v \in V$ is an arbitrary node, it is possible to construct a walk starting and ending at v in which each edge is traversed exactly once.*

Note that a *walk* is an alternating set of nodes and edges $v_0, e_1, v_1, e_2, \ldots, e_r, v_r$ where $e_i = (v_{i-1}, v_i) \in E$ for $i = 1, \ldots, r$.

For the analysis below, we need the following.

Proposition 12.3 *Given a complete graph H on node set V with edge lengths satisfying the triangle inequality, let $G = (V, E)$ be a connected Eulerian sub-*

graph of H. Then the original graph contains a Hamiltonian (STSP) tour of length at most $\sum_{e \in E} c_e$.

Proof. The proof is by construction. Suppose $m = |E|$ and $v = v_0, e_1, v_1, e_2, \ldots, e_m, v_m = v$ is a walk through the edges of G where each edge $e_1, \ldots, e_m$ occurs once and $e_i = (v_{i-1}, v_i)$ for $i = 1, \ldots, m$. Consider the list of nodes encountered in order $v_0, v_1, \ldots, v_m$. Suppose v_{i_k} is the kth distinct node to appear in the sequence. Then $v = v_{i_1}, v_{i_2}, \ldots, v_{i_n}, v_{i_{n+1}} = v$ is a tour. We now estimate its length. By the triangle property, the length of the subwalk between nodes v_{i_k} and $v_{i_{k+1}}$ is at least as long as the length of edge $f_k = (v_{i_k}, v_{i_{k+1}})$. More precisely $\sum_{j=i_k+1}^{i_{k+1}} c_{e_j} \geq c_{f_k}$. So the tour length $\sum_{k=1}^{n} c_{f_k} \leq \sum_{k=1}^{n} \sum_{j=i_k+1}^{i_{k+1}} c_{e_j} = \sum_{e \in E} c_e$. ∎

Now we can describe a first heuristic.

The Tree Heuristic for STSP

1. In the complete graph, find a minimum-length spanning tree with edges E_T and length $z_T = \sum_{e \in E_T} c_e$.
2. Double each edge of E_T to form a connected Eulerian graph.
3. Using Proposition 12.3, convert the Eulerian graph into a tour of length z^H.

Note that in Step 1 the degree of each node of the tree is nonzero. So when the edges are duplicated in Step 2, the degree of each node is even and positive.

Proposition 12.4 $\frac{z^H}{z} \leq 2$.

Proof. As every tour consists of a tree plus an additional edge, we obtain the lower bound $z_T \leq z$. By construction, the length of the Eulerian subgraph is $2z_T$. By the tour construction procedure, $z^H \leq 2z_T$ provides the upper bound. Thus we have $\frac{z^H}{2} \leq z_T \leq z \leq z^H$. ∎

Now we describe a second heuristic based on the construction of a shorter Eulerian subgraph. A matching is *perfect* if it is incident to every node of the graph.

The Tree/Matching Heuristic for STSP

1. In the complete graph, find a minimum-length spanning tree with edges E_T and length $z_T = \sum_{e \in E_T} c_e$.
2. Let V' be the set of nodes of odd degree in (V, E_T). Find a perfect matching M of minimum length z_M in the graph $G' = (V', E')$ induced on V' where E' is the set of edges of E_T with both endpoints in V'. $(V, E_T \cup M)$ is a connected Eulerian graph.
3. Using Proposition 12.3, convert the Eulerian graph into a tour of length z^C.

Note that here the perfect matching M has degree 1 for each node of V', and thus $(V, E_T \cup M)$ has positive even degree at every node.

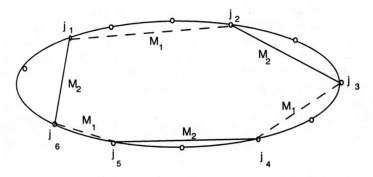

Fig. 12.3 Optimal tour longer than two matchings on V'

Proposition 12.5 $\frac{z^C}{z} \leq \frac{3}{2}$.

Proof. As above, $z_T \leq z$. Now suppose without loss of generality that the optimal tour of length z is the tour $1, 2, \ldots, n$. Let $j_1, j_2, \ldots, j_{2k}$ be the nodes of V' in increasing order. Consider the perfect matchings $M_1 = \{(j_1, j_2), (j_3, j_4), \ldots, (j_{2k-1}, j_{2k})\}$ of length z_{M_1} and $M_2 = \{(j_2, j_3), (j_4, j_5), \ldots, (j_{2k-2}, j_{2k-1}), (j_{2k}, j_1)\}$ of length z_{M_2}, both with endpoints V'. Again by the triangle inequality $c_{j_1, j_2} + c_{j_2, j_3} + \ldots + c_{j_{2k}, j_1} \leq \sum_{i=1}^{n-1} c_{i, i+1} + c_{n, 1} = z$ (see Figure 12.3). But now $z_M \leq z_{M_i}$ for $i = 1, 2$ and so $2z_M \leq z_{M_1} + z_{M_2} = c_{j_1, j_2} + c_{j_2, j_3} + \ldots + c_{j_{2k}, j_1} \leq z$. Finally, again using Proposition 12.3, $z^C \leq z_T + z_M \leq z + z/2 = (3/2)z$. ∎

12.5 MIP-BASED HEURISTICS

Below we discuss three heuristics for integer or mixed integer programs. The idea of the first is to take the linear programming solution at any node of the branch-and-bound tree and dive down the tree in the hope of finding a feasible solution. The other two can be implemented by anyone having access to a general mixed integer programming system.

A Dive-and-Fix Heuristic. We suppose that we have a mixed 0–1 problem. Given an linear programming solution (x^*, y^*), let $F = \{j : y_j^* \notin \{0, 1\}\}$ be the set of 0–1 variables that are fractional.

Initialization. Take the linear programming solution (x^*, y^*) at some node.

Basic Iteration:
As long as $F \neq \emptyset$,

Let $i = \arg\min_{j \in F}\{\min[y_j^*, 1 - y_j^*]\}$ (find the variable closest to integer).
If $y_i^* < 0.5$, fix $y_i = 0$ (if close to 0, fix to 0).
Otherwise set $y_i = 1$ (if close to 1, fix to 1).
Solve the resulting LP.
If the LP is infeasible, stop (the heuristic has failed).
Otherwise let (x^*, y^*) be the new linear programming solution.

Termination:
If $F = \emptyset$, (x^*, y^*) is a feasible mixed integer solution.

The next two heuristics use the mixed integer system in iterative mode, because the original problem is too large or difficult. For simplicity we decribe the heuristics for an IP, but the approach applies immediately to $MIPs$.

Suppose the problem can be written in the form:

$$(IP) \quad \begin{array}{l} z = \max c^1 x^1 + c^2 x^2 \\ A^1 x^1 + A^2 x^2 = b \\ x^1 \in Z_+^{n_1}, x^2 \in Z_+^{n_2}. \end{array}$$

We suppose that the variables x_j^1 for $j \in N_1$ are more important than the variables x_j^2 for $j \in N_2$ with $n_i = |N_i|$ for $i = 1, 2$. It may be that x^1 represent major investments, while x^2 represent less important decisions such as maintenance, or else the x^1 variables may represent decisions in the initial periods, while the x^2 variables represent decisions in later periods.

The idea is to solve two (or more) easier LPs or $MIPs$. The first one allows us to fix or limit the range of the more important x^1 variables, whereas the second allows us to choose good values for the variables x^2.

Relax-and-Fix Heuristic

1. *Relax.* Solve the relaxation

$$(MIP1) \quad \begin{array}{l} \bar{z} = \max c^1 x^1 + c^2 x^2 \\ A^1 x^1 + A^2 x^2 = b \\ x^1 \in Z_+^{n_1}, x^2 \in R_+^{n_2} \end{array}$$

in which the integrality of the x^2 variables is dropped. Let $(\bar{x}^1, \bar{x}^2)$ be the corresponding solution.

2. *Fix.* Fix the important variables x^1 at their values in $MIP1$, and solve the restriction

$$(IP2) \quad \begin{aligned} \underline{z} = \max c^1 x^1 + c^2 x^2 \\ A^1 x^1 + A^2 x^2 = b \\ x^1 = \overline{x}^1 \\ x^2 \in Z_+^{n_2}. \end{aligned}$$

Let $(\overline{x}^1, \tilde{x}^2)$ be the corresponding solution if $IP2$ is feasible.

3. *Heuristic Solution.* The heuristic solution is $x^H = (\overline{x}^1, \tilde{x}^2)$ with $\underline{z} = cx^H \leq z \leq \overline{z}$.

The idea of the next heuristic is similar, but here we suppose that an effective strong cutting plane algorithm is available, so that after adding cuts, at least some of the integer variables take values close to integer or to their final values. Again we suppose that the variables x^1 are in some way different from, or more important than, the x^2 variables.

Cut-and-Fix Heuristic

1. *Cut.* Apply a strong cutting plane algorithm to IP terminating with a tightened linear programming relaxation

$$(LP1) \quad \begin{aligned} \overline{z} = \max c^1 x^1 + c^2 x^2 \\ A^1 x^1 + A^2 x^2 = b \\ \tilde{A}^1 x^1 + \tilde{A}^2 x^2 \leq \tilde{b} \\ x^1 \in R_+^{n_1}, x^2 \in R_+^{n_2} \end{aligned}$$

with solution $(\overline{x}^1, \overline{x}^2)$. Here $\tilde{A}^1 x^1 + \tilde{A}^2 x^2 \leq \tilde{b}$ represent the cuts added to the initial linear programming relaxation.

2. *Fix (or Bound).* Choose ϵ. For $j \in N^1$, set $l_j = \lfloor \overline{x}_j^1 + \epsilon \rfloor$ and $u_j = \lceil \overline{x}_j^1 - \epsilon \rceil$. Solve the restriction

$$(IP2) \quad \begin{aligned} \underline{z} = \max c^1 x^1 + c^2 x^2 \\ A^1 x^1 + A^2 x^2 = b \\ \tilde{A}^1 x^1 + \tilde{A}^2 x^2 \leq \tilde{b} \\ l \leq x^1 \leq u \\ x^1 \in Z_+^{n_1}, x^2 \in Z_+^{n_2} \end{aligned}$$

with solution $(\tilde{x}^1, \tilde{x}^2)$ if $IP2$ is feasible.

3. *Heuristic Solution.* The heuristic solution is $x^H = (\tilde{x}^1, \tilde{x}^2)$ with $\underline{z} = cx^H \leq z \leq \overline{z}$.

Observe that if ϵ is small and positive, x^1 variables taking linear programming values within ϵ of integer in $LP1$ are fixed in $IP2$, while others are forced to take either the value $\lfloor \bar{x}_j^1 \rfloor$ or the value $\lceil \bar{x}_j^1 \rceil$. On the other hand, if ϵ is negative, all the x^1 variables can still take at least two values in $IP2$.

12.6 NOTES

For a simple treatment of the topics in Sections 12.2 and 12.3, see [Ree93]. A highly readable survey is [Pir96]. See also the annotated bibliography on local search [AarVer97]. At a more advanced level the recent book [AarLen97] is devoted to local search and its extensions, including chapters written by specialists on the complexity of local search, simulated annealing, tabu search, genetic algorithms, applications to vehicle routing, *TSP*, machine scheduling VLSI layout, and so forth. An issue of *Management Science* [FisRin88] is dedicated to the subject of heuristics.

12.2 Greedy heuristics are part of the folklore; see for instance [KueHam63] for an early application to location problems. Local search heuristics date at least to [Cro58], [ReiShe65].

12.3 Tabu search and simulated annealing have been applied successfully to find good quality feasible solutions to a remarkably wide range of problems. Tabu search started with the work of [Glo86], [Glo89], [Glo90]. The origins of simulated annealing heuristics are attributed to [Metetal53], [Kiretal83], [Cern85]. Theoretical results, relying on the asymptotic behavior of Markov chains, can be used to show that the simulated annealing algorithm almost surely terminates with a global optimum if run with a slow cooling schedule for a long enough time. However, these results are inapplicable in practice, and fail to explain the many successes of this method. A recent bibliography of simulated annealing and tabu search is [OsmLap96]. Genetic algorithms originated with the work of [Hol75], [Gol89]. The generalized assignment problem is treated in [Beas97].

Another recent approach is that of neural networks (see the texts cited above). Constraint logic programming provides an alternative approach for certain discrete problems.

12.4 The first worst-case analysis of a heuristic for a scheduling problem is in [Gra66], but the analysis of bin packing heuristics [Johetal74] really seems to have initiated much of the work in this area. The tree/matching heuristic for $STSP$ is from [Chr76]; see also [CorNem78].

For certain problems it has been shown that finding a heuristic giving a performance guarantee of α or better is only possible if $\mathcal{P} = \mathcal{NP}$, while for others one can find so-called *fully polynomial approximation schemes* with performance guarantees of α for any α in a time polynomial in the input length and $1/\alpha$. The book [Hoc95] includes surveys on approximation results for bin packing, covering and packing, and network design problems among others, as well as on randomized algorithms and on the hardness of approximation. Another recent survey is [Shm95].

12.5 Variants of the dive-and-fix heuristic are common; see [NemSavSig94]. The relax-and-fix heuristic is also well known. Two recent examples of the relax-and-fix and cut-and-fix heuristics are in [BieGun98] and [Belvetal98] respectively.

12.7 EXERCISES

1. Apply the different heuristics presented to an instance of $STSP$ with the following distance matrix:

$$\begin{pmatrix} - & & & & & & \\ 28 & - & & & & & \\ 57 & 28 & - & & & & \\ 72 & 45 & 20 & - & & & \\ 81 & 54 & 3 & 10 & - & & \\ 85 & 57 & 28 & 20 & 22 & - & \\ 80 & 63 & 57 & 72 & 81 & 63 & - \end{pmatrix}.$$

Devise at least one new heuristic and apply it to this instance.

2. Apply greedy and local neighborhood heuristics to an instance of the problem of most profitably allocating clients to at most K depots:

$$\max \sum_{i \in M} \sum_{j \in N} c_{ij} x_{ij}$$
$$\sum_{j \in N} x_{ij} = 1 \text{ for } i \in M$$
$$x_{ij} \leq y_j \text{ for } i \in M, j \in N$$
$$\sum_{j \in N} y_j \leq K$$
$$x \in B^{mn}, y \in B^n,$$

with $m = 7$ clients, $n = 6$ potential depots, $K = 3$, and

$$(c_{ij}) = \begin{pmatrix} 2 & 3 & 7 & 3 & 6 & 1 \\ 3 & 1 & 1 & 8 & 10 & 4 \\ 6 & 2 & 3 & 1 & 2 & 7 \\ 8 & 1 & 4 & 6 & 2 & 3 \\ 4 & 4 & 3 & 3 & 4 & 3 \\ 2 & 8 & 3 & 6 & 3 & 2 \\ 6 & 5 & 3 & 2 & 7 & 4 \end{pmatrix}.$$

3. Devise a greedy heuristic for the set covering problem.

4. Consider the problem of finding a maximum cardinality matching from Chapter 4. A matching $M \subseteq E$ is *maximal* if $M \cup \{f\}$ is not a matching for

any $f \in E \setminus M$. Let z be the size of a maximum matching, and z^H be the size of a *maximal* matching. Show that $z^H \geq \frac{1}{2}z$.

5. Consider the 0–1 knapsack problem: $z = \max\{\sum_{j \in N} c_j x_j : \sum_{j \in N} a_j x_j \leq b, x \in B^n\}$ with $a_j > 0$ for $j \in N$. Consider a greedy heuristic that chooses the better of the integer round down of the linear programming solution, and the best solution in which just one variable is set to one. Show that $z^G \geq \frac{1}{2}z$.

6. Show that the 0–1 covering problem

$$z = \min \sum_{j=1}^n f_j x_j$$
$$\sum_{j=1}^n a_{ij} x_j \geq b_i \text{ for } i \in M$$
$$x \in B^n$$

with $a_{ij} \geq 0$ for $i \in M, j \in N$ can be written in the form

$$z = \min\{\sum_{j \in S} f_j : g(S) = g(N)\}$$

where $g : \mathcal{P}(N) \to R^1_+$ is a nondecreasing set function. What is g? Show that $g(S) + g(T) \geq g(S \cup T) + g(S \cap T)$ for all $S, T \subseteq N$.

7. Consider the generalized transportation problem:

$$z = \min \sum_{i=1}^m \sum_{j=1}^n c_{ij} x_{ij}$$
$$\sum_{j=1}^n x_{ij} \leq a_i \text{ for } i = 1, \ldots, m$$
$$\sum_{i=1}^m C_i x_{ij} \geq d_j \text{ for } j = 1, \ldots, n$$
$$x \in Z^{mn}_+.$$

(i) Propose a heuristic for this problem.
(ii) Propose an exact algorithm.

8. Devise a heuristic for the instance of GAP in Exercise 10.4.

13
From Theory to Solutions

13.1 INTRODUCTION

The aim in this final chapter is to indicate how the ideas presented earlier can be put to use to improve the formulation and solution of integer programs. Specifically we discuss briefly the type of software available, then we ask *how* an integer programmer might look for a new valid inequality, or an improved formulation for a specific problem, and finally we look at two practical applications and ask *how* an integer programmer might try to solve them. The goal is not to study any problem exhaustively, nor to present computational results for different approaches, but rather to indicate the questions to be asked and the steps to be considered in trying to produce results for a particular problem.

13.2 SOFTWARE FOR SOLVING INTEGER PROGRAMS

Two essential tools for an integer programmer are a modeling language, and a mixed integer programming system.

A *modeling language* provides a way for the user to write out a *model* of his mixed integer programming formulation in a formal way that closely resembles the mathematical formulations used throughout this book. In addition such languages allow recuperation of the data from files or databases, and calculations based on the data. Given the model written in the modeling language as input, the output is typically a representation of the mixed integer program in a standard (but unfortunately highly unreadable) format, known as *MPS format*.

There are at least two crucial advantages in using a modeling language. First, the model provides documentation, making it relatively easy to return to a model several weeks or months later. Second, with such a model, changes to a formulation can be made very easily and quickly. The modified model can then be used to regenerate an MPS file, and the modified formulation can often be resolved within seconds or minutes. An example of a model in a representative language is shown below. This formulation and instance are discussed in Section 13.4.

```
!----------------------------------------------------------------
MODEL FCNF ! Fixed Charge Network Flow Model
           ! Instance from Section 13.4
           ! Modelling language: XPRESS MP-MODEL
! Parameters/Indices
LET N=4 ! Number of nodes
LET M=8 ! Number of arcs
! Vectors/Matrices
TABLES
B(N) ! Demand at node i
TAIL(M) ! Tail of arc e
HEAD(M) ! Head of arc e
C(M) ! Fixed cost of arc e
!----------------------------------------------------------------
! Usually read from files
DATA
B(1)=-6,2,3,1
TAIL(1)=1,1,4,1,2,3,4,3
HEAD(1)=4,2,2,3,3,2,3,4
C(1)=5,2,7,3,2,4,9,12
!----------------------------------------------------------------
VARIABLES
x(M) ! Flow x(e) in arc e for e=1,...,m
y(M) ! Binary variable y(e)=1 if arc is open

CONSTRAINTS
!      Flow conservation at node i
BAL(i=1:N): SUM(e=1:M | HEAD(e) .eq. i)x(e)                    &
            -SUM(e=1:M | TAIL(e) .eq. i)x(e) = B(i)
! Variable upper bound constraint on arc e
VUB(e=1:M): x(e) < (-B(1))*y(e)
! Fixed cost of installing arcs
MINOBJ: SUM(e=1:M)C(e)*y(e) $
```

```
BOUNDS
y(e=1:M)   .bv.  ! Binary Variable
!------------------------------------------------
GENERATE   ! Produce the MPS file
!------------------------------------------------
! Usually further instructions for presenting
!     the solution in readable form
END
!------------------------------------------------
```

A *mixed integer programming system* is a program that reads in an mixed integer program in MPS format and possibly other formats, creates an internal representation of the problem, referred to as the *matrix*, and then typically attempts to solve it by linear-programming-based branch-and-bound as described in Chapter 7. Recently some of the major systems have started using branch-and-cut, generating 0–1 knapsack inequalities and in some cases mixed integer inequalities as well.

Certain systems also provide *optimization subroutine libraries*, permitting the user to build his own special-purpose system. These libraries contain subroutines to load and unload *LP*s and *MIP*s, to optimize matrices, to retrieve data from matrices, to modify matrices including the addition or deletion of constraints and/or columns, to examine both primal and dual solutions, and so on. Many special purpose branch-and-cut and column generation algorithms have been developed in this way.

One of the most recent developments is an extended modeling and optimization library that integrates a modeling language and a mixed integer programming system. Thus the user can simultaneously have access to the model file and the problem matrix, allowing for the higher-level development of new iterative optimization strategies for very large models, including relax-and-fix heuristics, specially adapted branching strategies, column generation, and so on.

13.3 HOW DO WE FIND AN IMPROVED FORMULATION?

We first examine two simple lot-sizing models that appear as submodels in the production planning application to be examined in Section 13.5. The two models, both of which have been discussed earlier, differ in their complexity.

13.3.1 Uncapacitated Lot-Sizing

Consider the set $X^{ULS} =$

$$\{(x,s,y) \in R_+^n \times R_+^n \times B^n : s_{t-1}+x_t = d_t+s_t, x_t \leq (\sum_{i=1}^{t} d_i)y_t \text{ for } t=1,\ldots,n\}$$

224   FROM THEORY TO SOLUTIONS

We typically ask a series of questions.

Q1. Is optimization over X^{ULS} polynomially solvable?

A1. Yes. This is shown in Section 5.2. Therefore, as indicated in Chapters 2 and 6, there is some hope of finding an explicit description of conv(X^{ULS}).

One way to improve a formulation is to add valid inequalities. To make a start at finding a valid inequality, one approach is to look at the non-integral solutions obtained when solving the linear programming relaxation of $\min\{px + hs + fy : (x, s, y) \in X^{ULS}\}$.

We consider an instance with $d = (6, 7, 4, 6, 3, 8), h = (1, 1, 3, 1, 1, 2)$ and $f = (8, 12, 8, 6, 10, 23)$. As $d_1 > 0$, we immediately add $y_1 = 1$ to the formulation.

The resulting linear programming solution (x^1, s^1, y^1) is shown in Figure 13.1.

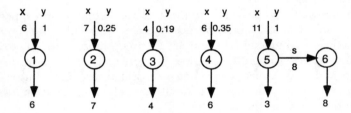

Fig. 13.1   First ULS solution

Q2. How can the point (x^1, s^1, y^1) be cut off?

Observation 13.1. As in the development of the dynamic programming recursion for ULS, the solution decomposes between successive points with $s_t = 0$. So it suffices to look at the solution lying between $s_1^1 = 0$ and $s_2^1 = 0$, namely $x_2^1 = 7, y_2^1 = 0.25, s_1^1 = 0, s_2^1 = 0$ for which the y variable is not integer.

A2a. Try to use known cutting planes. Consider the set

$$\{(x_2, y_2, s_1, s_2) \in R_+^1 \times B^1 \times R_+^2 : s_1 + x_2 = d_2 + s_2, x_2 \leq (\sum_{i=2}^{n} d_i) y_2)\}.$$

The flow cover inequality with variable x_2 in the cover C and excess $\lambda = \sum_{i=2}^{n} d_i - d_2$ gives the inequality $x_2 + d_2(1 - y_2) \leq d_2 + s_2$, or in general

$$x_t \leq d_t y_t + s_t \text{ for } t = 1, \ldots, n.$$

A2b. Try to find a valid inequality directly. Again just consider the partial solution $x_2^1 = 7, y_2^1 = 0.25, s_1^1 = 0, s_2^1 = 0$. Logically in any feasible solution to ULS with $y_2 = 0$, there is no production in period 2, and so the demand $d_2 = 7$ for period 2 must be contained in the entering stock s_1. Thus $s_1 \geq 7$ if $y_2 = 0$. Now we try to convert this observation into a valid inequality.

The inequality $s_1 \geq d_2(1 - y_2)$ does the trick when $y_2 = 0$. So we just need to check that it is also valid when $y_2 = 1$. But in this case the inequality reduces to $s_1 \geq 0$, which is a constraint of the initial problem. So in general we have a valid inequality

$$s_t \geq d_t(1 - y_t) \text{ for } t = 1, \ldots, n.$$

Using the equality $s_{t-1} + x_t = d_t + s_t$, it is readily checked that the two inequalities we have come up with are the same, and those for $t = 2, 3, 4$ cut off the linear programming solution (x^1, s^1, y^1).

To go a step further, we add these inequalities to the formulation, and the new linear programming solution is as shown in Figure 13.2.

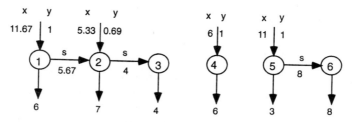

Fig. 13.2 Second ULS solution

Again the question is how to cut off the point (x^2, s^2, y^2). As $s_3^2 = 0$, it suffices to look at the nonintegral solution for periods 1–3.

A2c. Extending the argument in A2b, we observe that if there is no production in periods 2 and 3, then $s_1 \geq d_2 + d_3 = 11$. This immediately gives the family of valid inequalities

$$s_{k-1} \geq (\sum_{t=k}^{l} d_t)(1 - y_k - \ldots - y_l) \text{ for } 1 \leq k \leq l \leq n$$

and among them $s_1 \geq 11(1 - y_2 - y_3)$ cuts off (x^2, s^2, y^2).

Again we add these inequalities to the formulation, and the new linear programming solution is as shown in Figure 13.3.

Here we observe that the cuts added in the last two iterations are both satisfied at equality. Somehow we need something stronger.

A2d. We have used the two valid inequalities $s_1 \geq d_2(1 - y_2)$ and $s_1 \geq (d_2 + d_3)(1 - y_2 - y_3)$. Comparing the two, we see that the d_2 term in the second inequality is weaker than in the first. The valid inequality

$$s_1 \geq d_2(1 - y_2) + d_3(1 - y_2 - y_3)$$

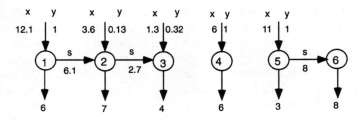

Fig. 13.3 Third ULS solution

takes this observation into account, and has exactly the right condition needed to include the demands d_2, or $d_2 + d_3$. After adding this inequality, we obtain a solution with y integer. The general form of this last inequality is

$$s_{k-1} \geq \sum_{t=k}^{l} d_t(1 - y_k - \ldots - y_t) \text{ for } 1 \leq k \leq l \leq n. \tag{13.1}$$

Before leaving this example, there are still several questions that we might ask about the use and strength of the inequalities we have found.

Q3. Are the inequalities (13.1) as strong as possible (facet-defining), or are they nonnegative combinations of other valid inequalities?

A3. Yes. The inequalities are facet-defining, see Exercise 13.5.

Q4. Do the constraints of the original formulation plus the $O(n^2)$ inequalities (13.1) describe conv(X^{ULS})?

A4. No. The inequality $x_k \leq (\sum_{t=k}^{l} d_t)y_j + s_l$ is valid and facet-defining for any $l \geq k$, and is not equivalent to any inequality of the form (13.1) when $l > k$. However, the inequalities (13.1) are sufficient to solve most practical instances.

Q5. Is the separation problem for the inequalities (13.1) and a point (x^*, s^*, y^*) easy?

A5. Yes. Fix k, and find the smallest value $l \geq k$ for which $y_k^* + \ldots + y_l^* \geq 1$. If $l > k$, check if

$$s_{k-1}^* < \sum_{t=k}^{l-1} d_t(1 - y_k^* - \ldots - y_t^*).$$

If so, this is the most violated inequality for the given value of k. If not, no such inequality is violated.

In conclusion, one has the option of either reformulating by adding the $O(n^2)$ inequalities a priori, or of using a simple separation routine to generate the inequalities (13.1) as cuts when they are violated.

Q6. Is there an extended formulation for ULS?

A6. There are several including the one presented in Section 1.6; see also Subsection 9.2.3. However, in spite of their strength, most such formulations have the disadvantage of having an order of magnitude more variables and constraints.

13.3.2 Capacitated Lot-Sizing

Here we consider the more constrained set $X^{CLS} =$

$\{(x,s,y) \in R_+^n \times R_+^n \times B^n : s_{t-1} + x_t = d_t + s_t, x_t \leq C_t y_t \text{ for } t = 1,\ldots,n\}.$

We start with the same questions.

Q1. Is optimization over X^{CLS} polynomially solvable?

A1. No. CLS is shown to be $\mathcal{NP}$-hard in Section 6.3. Therefore the best we can hope for is to find a good approximation of $\text{conv}(X^{CLS})$.

Q2. How do we find valid inequalities?

We again try to make use of existing inequalities, by considering relaxations for which valid inequalities are known. After looking at single periods, it is natural to aggregate together flow conservation equations from consecutive periods.

A2a. Consider the relaxation

$$\{(s_{k-1}, y_k, \ldots, y_l) \in R_+^1 \times B^{l-k+1} : s_{k-1} + \sum_{j=k}^{l} C_j y_j \geq \sum_{j=k}^{l} d_j\},$$

obtained by summing the constraints $s_{t-1} + x_t = d_t + s_t$ and replacing x_t by its upper bound $C_t y_t$ for periods $k, \ldots, l$, and by replacing s_l by its lower bound of 0.

One can either attempt to generate an MIR inequality off this constraint (see Section 8.7), or else one can view it as a 0-1 knapsack problem with a continuous variable s_{k-1}. Using the approach used in Section 9.3, one can fix $s_{k-1} = 0$, apply a knapsack separation heuristic, and then lift back in the continuous variable.

A2b. Aggregating the same constraints as above, replacing s_k by its lower bound 0, and adding one of the valid inequalities for ULS (see A4 of Subsection 13.2.1) leads to the set

$$\{(x_k,\ldots,x_l,s_l,y_k,\ldots,y_l) \in R_+^{l-k+1} \times R_+^1 \times B^{l-k+1} :$$

$$\sum_{j=k}^{l} x_j \leq \sum_{j=k}^{l} d_j + s_l,$$

$$x_j \leq C_j y_j, x_j \leq (\sum_{t=j}^{l} d_t) y_j + s_l \text{ for } j = k,\ldots,l\}.$$

Fixing $s_l = 0$, this becomes a single node flow set (see Section 9.4), for which one can generate flow cover inequalities before reintroducing s_l.

Consider the solution shown in Figure 13.4 for an instance with $n = 6, d = (6, 7, 4, 6, 3, 8)$ and $C = (15, 8, 10, 10, 5, 10)$.

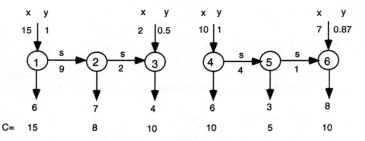

Fig. 13.4 CLS solution

Again because $s_3 = 0$, we can decompose the solution into two parts, one for periods 1-3 and the other for periods 4-6.

For periods 1-3 we look at the relaxation suggested in A2a, giving the set

$$\{y \in B^3 : 15y_1 + 8y_2 + 10y_3 \geq 17\}$$

and the fractional point $(y_1, y_2, y_3) = (1, 0, \frac{1}{2})$. The 0–1 knapsack separation routine, or the Chvátal-Gomory procedure for $y_1 + \frac{8}{15}y_2 + \frac{10}{15}y_3 \geq \frac{17}{15}$ both lead to the violated valid inequality

$$y_1 + y_2 + y_3 \geq 2.$$

For periods 4-6 we look at the relaxation suggested in A2b:

$$(x_4, x_5, x_6, y_4, y_5, y_6, s_6) \in R_+^3 \times B^3 \times R_+^1 : x_4 + x_5 + x_6 \leq 17 + s_6,$$

$$x_4 \leq \min(10, 17)y_4 + s_6, x_5 \leq \min(5, 11,)y_5 + s_6, x_6 \leq \min(10, 8)y_6 + s_6$$

with fractional point $(x_4, x_5, x_6, y_4, y_5, y_6, s_6) = (10, 0, 7, 1, 0, 0.875, 0)$.

The flow cover inequality with cover variables (x_4, x_6) and $\lambda = 10+8-17 = 1$ leads to the violated valid inequality

$$x_4 + x_6 + 9(1 - y_4) + 7(1 - y_6) \leq 17 + s_6.$$

Q3. Are the inequalities obtained as tight as possible?

A3. In practice it may not be important to check that the inequalities define facets. However, it is worth checking whether the inequalities can be easily strengthened, or possibly decomposed into two or more valid inequalities.

Q4. Is the separation problem for the inequalities arising in A2a or A2b easy?

A4. In practice, a separation heuristic, such as that for flow covers, can be devised.

13.4 FIXED CHARGE NETWORKS: REFORMULATIONS

13.4.1 The Single Source Fixed Charge Network Flow Problem

The basic *fixed charge network flow problem* (*FCNF*) involves a digraph $D = (N, A)$, a demand vector $b \in R^n$, capacity and cost vectors $u, c \in R^m$. The problem is to choose a set $A' \subseteq A$ of arcs of minimum cost such that there is a feasible flow in the resulting digraph $D' = (N, A')$ satisfying the demand and capacity constraints. We restrict our attention to *single-source* problems, that is, we suppose that node 1 is the *root* or source at which flow enters, so $b_1 < 0$, $b_i > 0$ for $i \in T \subseteq N \setminus \{1\}$, and $b_i = 0$ otherwise. Note that a necessary condition for feasibility is that $b_1 = -\sum_{i \in N \setminus \{1\}} b_i$. We also assume that $c_{ij} > 0$ for all $(i,j) \in A$, and set $n = |N|$ and $m = |A|$.

To formulate this problem, we define variables

x_{ij} to be the flow in arc $(i,j) \in A$, and
$y_{ij} = 1$ if arc $(i,j) \in A$ is open, and $y_{ij} = 0$ otherwise.

We then obtain the formulation

$$\min \sum_{(i,j) \in A} c_{ij} y_{ij} \tag{13.2}$$

$$\sum_{j \in V^-(i)} x_{ji} - \sum_{j \in V^+(i)} x_{ij} = b_i \text{ for } i \in N \tag{13.3}$$

$$0 \leq x_{ij} \leq u_{ij} y_{ij} \text{ for } (i,j) \in A \tag{13.4}$$

$$y \in B^m. \tag{13.5}$$

Small examples suffice to show that the formulation (13.2)–(13.5) is not very strong, and large instances cannot be solved with this formulation, so here we consider how alternative formulations might be found for $FCNF$, as well as for a special case of $FCNF$.

To obtain some initial ideas, we again look at the linear programming solution obtained for a small instance. With $n = 4, m = 8$, and the network shown in Figure 13.5, the solution of the linear programming relaxation is $x_1 = 1, y_1 = \frac{1}{6}, x_2 = 2, y_2 = \frac{1}{2}, x_4 = 3, y_4 = \frac{1}{2}, x_i = y_i = 0$ otherwise. The data is given in Section 13.1 where this instance is presented as a example of a formulation written in a modeling language.

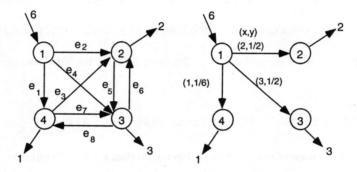

Fig. 13.5 Fixed charge network flow

Observation 13.2. The formulation is unlikely to be effective because in the linear programming solution $y_{ij} = x_{ij}/u_{ij}$ for all $(i,j) \in A$. Therefore for an uncapacitated problem with u_{ij} large, the fixed arc costs are severely underestimated and so the linear programming bound is particularly weak.

Observation 13.3. Consider the flow into node 4. Even though all the demand for the node flows in the arc $(1,4)$, $y_{14} = \frac{1}{6}$ is far from taking the value 1 that it should have in an integer programming solution with $x_{14} > 0$.

Observation 13.4. Consider node 4 again. As $b_4 = 1$, there must be some flow entering the node, and therefore at least one arc entering the node must be open.

A Multicommodity Reformulation. Observation 13.3 leads to the idea of looking for a formulation in which the destination of the flow in each arc is explicit. Specifically if we introduce new multicommodity variables

z_{ij}^k is the flow of commodity k in arc (i,j) whose destination is node k, we have that $z_{ij}^k \leq b_k$, and $z_{ij}^k = 0$ if $y_{ij} = 0$, leading to the variable upper bound constraint $z_{ij}^k \leq b_k y_{ij}$. Writing a formulation with these new distinct destination or *multicommodity* variables is straightforward.

$$\min \sum_{(i,j) \in A} c_{ij} y_{ij} \tag{13.6}$$
$$\sum_{j \in V^-(i)} z_{ji}^k - \sum_{j \in V^+(i)} z_{ij}^k = d_i^k \text{ for } i \in N, k \in T \tag{13.7}$$
$$0 \leq z_{ij}^k \leq \min(u_{ij}, b_k) y_{ij} \text{ for } (i,j) \in A, k \in T \tag{13.8}$$
$$\sum_{k \in T} z_{ij}^k = x_{ij} \text{ for } (i,j) \in A \tag{13.9}$$
$$x_{ij} \leq u_{ij} \text{ for } (i,j) \in A \tag{13.10}$$
$$y \in B^m. \tag{13.11}$$

where $d_i^k = -b_k$ for $i = 1$, $d_i^k = b_k$ for $i = k$ and $d_i^k = 0$ otherwise.

Note now that if $z_{ij}^k = b_k$ in a linear programming solution, this constraint will force $y_{ij} = 1$. In particular, in our example if $z_{14}^4 = 1$, then $y_{14} = 1$. Thus we have obtained an apparently stronger formulation, but at the cost of an order of magnitude more variables and constraints.

13.4.2 The Directed Subtree Problem

When the single-source $FCNF$ is uncapacitated, for instance when $u_{ij} \geq -b_1$ for all $(i,j) \in A$, then the resulting digraph $D' = (N, A')$ is feasible if and only if there is a directed path from node 1 to every node $i \in T$ with a positive demand. If such a digraph contains no cycles, it is a *directed subtree* with root $r = 1$. Now in the original formulation of $FCNF$, it suffices to take $b_1 = -|T|$ and $b_i = 1$ for $i \in T$, and in the multicommodity reformulation the constraints (13.9) and (13.10) involving x_{ij} are redundant, and so we obtain

$$\min \sum_{(i,j) \in A} c_{ij} y_{ij} \tag{13.12}$$
$$\sum_{j \in V^-(i)} z_{ji}^k - \sum_{j \in V^+(i)} z_{ij}^k = d_i^k \text{ for } i \in N, k \in T \tag{13.13}$$
$$0 \leq z_{ij}^k \leq y_{ij} \text{ for } (i,j) \in A, k \in T \tag{13.14}$$
$$y \in B^m. \tag{13.15}$$

where $d_i^k = -1$ for $i = 1$, $d_i^k = 1$ for $i = k$ and $d_i^k = 0$ otherwise.

Another possibility is to look for a formulation involving only the y_{ij} variables for $(i,j) \in A$. Observation 13.4 immediately suggests the valid inequality

$$\sum_{j \in V^-(i)} y_{ji} \geq 1$$

for all $i \in T$.

Q1. Can this inequality be generalized?

A1. As there is a positive flow from node 1 to node $i \in T$, there must be a positive flow across every $(1,i)$ cut-set separating 1 from i. This means that there must be at least one arc directed across each of these cut-sets, leading to the larger family of valid inequalities

$$\sum_{(ij)\in\delta(X,\bar{X})} y_{ij} \geq 1$$

for all $X \subset V$ with $1 \in X$ and $T \cap \bar{X} \neq \emptyset$.

Q2. Do these new inequalities provide a valid formulation of the directed tree problem, namely is the *dicut formulation*

$$\min \sum_{(i,j)\in A} c_{ij} y_{ij} \qquad (13.16)$$

$$\sum_{(ij)\in\delta(X,\bar{X})} y_{ij} \geq 1 \text{ for } X \subset V \text{ with } 1 \in X, T \cap \bar{X} \neq \emptyset \qquad (13.17)$$

$$y \in B^m \qquad (13.18)$$

a valid formulation?

A2. Yes. For any $i \in T$, if every cut-set separating 1 and i contains an arc, then there exists a directed path from 1 to i, and thus a minimal feasible solution is the incidence vector of a directed subtree. Conversely the incidence vector of any directed subtree is feasible.

Q3. Can anything be said about the strength of the multicommodity (13.12)–(13.15) and dicut formulations (13.16)–(13.18) for the directed subtree problem?

A3. We know from Section 1.7 that formulations can be compared by examining their projections in the original space of variables. For a given $y \in R^m_+$ with $y_{ij} \leq 1$ for all $(i,j) \in A$, the linear programming relaxation of the multicommodity formulation is feasible if and only if for each $i \in T$, there is a feasible flow of one unit from node 1 to node i in the digraph with arc capacities y_{ij}. By the max-flow/min cut theorem (Section 3.4), this holds if and only if the capacity of every $1-i$ cut is at least 1. But this is precisely the condition for y to be feasible in the linear programming relaxation of the dicut formulation. Thus we conclude that the strength of the two formulations is the same.

13.5 MULTI-ITEM SINGLE MACHINE LOT-SIZING

A set of products or items have to be produced on a single machine. Because of important start-up costs and changeover times between the production of different items, the planning horizon has been broken up into equally spaced time intervals (periods such as a day, or an 8-hour shift) such that only one item is produced per period. Also certain items have minimum run-times, and others have minimum down-times specified between successive production runs of the item.

The data consists of

- the demand d_t^i for item i in period t
- the storage cost h^i per period per unit of item i
- the start-up cost g^i for item i
- the changeover cost q^{ij} when passing from the production of i to j
- the changeover time Δ^{ij} when passing from the production of i to j
- the production rate ρ^i per period for item i
- the minimum run-time α^i for item i
- the minimum down-time β^i for item i
- the production capacity C_t per period

For simplicity we assume that the data have been preprocessed so that the initial stocks and safety stocks (lower bounds on stocks) are zero. In addition we assume that the basic unit of each item i is the amount produced per unit of time. With this assumption $\rho^i = 1$ for all $i = 1, \ldots, m$, and $C_t = C$ is the length of a period.

We make the obvious choice of variables:

x_t^i the amount of time that item i is produced in period t
s_t^i the stock of i at the end of t measured in time units
$y_t^i = 1$ if the machine is set up for i in t
$z_t^i = 1$ if a production run of i starts in t
$\chi_t^{ij} = 1$ if item i is produced in period $t-1$ and j in t.

This leads to the formulation

$$\min \sum_{i=1}^{m} \sum_{t=1}^{n} (h^i s_t^i + g^i z_t^i + \sum_{j:j \neq i} q^{ij} \chi_t^{ij}) \qquad (13.19)$$
$$s_{t-1}^i + x_t^i = d_t^i + s_t^i \text{ for all } i,t \qquad (13.20)$$
$$x_t^i \leq C y_t^i - \sum_{j:j \neq i} \Delta^{ji} \chi_t^{ji} \text{ for all } i,t \qquad (13.21)$$
$$z_t^i \geq y_t^i - y_{t-1}^i \text{ for all } i,t \qquad (13.22)$$
$$z_t^i \leq y_t^i \text{ for all } i,t \qquad (13.23)$$
$$z_t^i \leq 1 - y_{t-1}^i \text{ for all } i,t \qquad (13.24)$$
$$\chi_t^{ij} \geq y_{t-1}^i + y_t^j - 1 \text{ for all } i,j,t \text{ with } i \neq j \qquad (13.25)$$
$$\chi_t^{ij} \leq y_{t-1}^i \text{ for all } i,j,t \text{ with } i \neq j \qquad (13.26)$$
$$\chi_t^{ij} \leq y_t^j \text{ for all } i,j,t \text{ with } i \neq j \qquad (13.27)$$

$$z_t^i \leq y_{t+u-1}^i \text{ for all } i, t, u = 1, \ldots, \alpha^i \tag{13.28}$$
$$z_t^i \leq 1 - y_{t-u}^i \text{ for all } i, t, u = 1, \ldots, \beta^i \tag{13.29}$$
$$\sum_{i=1}^m y_t^i = 1 \text{ for all } t \tag{13.30}$$
$$s_t^i, x_t^i \geq 0, y_t^i, z_t^i \in \{0, 1\} \text{ for all } i, t \tag{13.31}$$
$$\chi_t^{ij} \in \{0, 1\} \text{ for all } i, j, t \text{ with } i \neq j \tag{13.32}$$

where (13.20) are flow conservation constraints, (13.21) are production capacity constraints, (13.22)–(13.24) define z_t^i as start-up variables, (13.25)–(13.27) define χ_t^{ij} as changeover variables, (13.28), (13.29) express the minimum run time and minimum down-time conditions, and (13.30) the single item per period restriction. For simplicity we ignore the fact that zero production is allowed even when an item is set up for a production run of one or several periods.

We suppose that this application is to be solved with a standard mixed integer programming system, and so no significant change of variables is envisaged. Thus the available options are to strengthen constraints, to add valid inequalities, plus possibly knapsack and mixed integer cuts that can be generated automatically, or to use an extended formulation involving new variables.

We examine in turn various possibilities.

Improving Branch-and-Bound Performance. We present three simple observations that may radically modify the performance of a branch-and-bound algorithm.

Observation 13.5. If the variables y_t^i are all 0–1 valued and constraints (13.22)–(13.27) are present, the variables z_t^i and χ_t^{ij} will automatically take 0–1 values. Therefore it may be possible to make the z and χ variables continuous.

Observation 13.6. If the start-up costs $\{g^i\}$ and changeover costs $\{q^{ij}\}$ are positive, one might consider whether it is valid and computationally interesting to drop constraints (13.23),(13.24),(13.26),(13.27) for the products without minimum run and down times, on the grounds that z_t^i and χ_t^{ij} will only take the value 1 if forced. Note also that (13.23) reappears as one of the constraints (13.28), and (13.24) as one of the constraints of (13.29).

Observation 13.7. Giving priorities to the binary variables y_t^i based on the period t, and/or the importance of the product i should be considered.

Problem-Specific Cutting Planes. Note first that with the relaxation $x_t^i \leq Cy_t^i$ of (13.21), the problem contains capacitated lot-sizing (CLS) relaxations for each item i, and thus the inequalities for both ULS and CLS derived in Section 13.2 can be used.

It is also natural to ask whether the basic lot-sizing inequalities can be improved in the presence of the start-up and/or changeover variables. One improvement is easily seen. The inequalities

$$s_{k-1} \geq \sum_{t=k}^{l} d_t(1 - y_k - \ldots - y_t)$$

for ULS were valid because no production between periods k and t, which can be restated as $y_k + \ldots + y_t = 0$, implies that the demand d_t must be included in the entering stock s_{k-1}. With start-up variables, the condition $y_k + z_{k+1} + \ldots + z_t = 0$ also implies no production between periods k and t, and thus demonstrates the validity of

$$s_{k-1} \geq \sum_{t=k}^{l} d_t(1 - y_k - z_{k+1} - \ldots - z_t) \text{ for } 1 \leq k \leq l \leq n.$$

This dominates the previous inequality as $z_i \leq y_i$ for $i = 1, \ldots, n$.

Modeling Minimum Run-Times. If an item must be produced during at least α periods, there can be at most one start-up of the item in the interval $[t - \alpha + 1, t]$. In addition if it starts at some time in this interval, then the machine must still be set up in period t. This shows that the inequality

$$\sum_{u=1}^{\alpha} z_{t-u+1} \leq y_t \text{ for } 1 \leq t \leq n$$

is valid for each item, and can be used to replace the inequalities (13.23) and (13.28).

Similarly for the down times, the inequality

$$\sum_{u=1}^{\beta} z_{t+u} \leq 1 - y_t \text{ for } 1 \leq t \leq n$$

is valid, and can replace (13.24) and (13.29).

Modeling Start-ups and Changeovers. Here we consider the constraints (13.22)–(13.27) and (13.30) linking the y, z, and χ variables. First we add a small number of additional variables:

$\chi_t^{jj} = 1$ if item j is produced in periods $t - 1$ and t.

Observe that if we know for which item the machine is setup in period 1, the χ variables tell us everything about the state of the machine from then on. More specifically they determine the path taken through the acyclic network shown in Figure 13.6. The path shown corresponds to the solution with $y_1^1 = y_2^2 = y_3^2 = y_4^1 = y_5^3 = 1$.

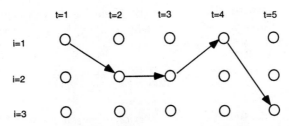

Fig. 13.6 Representation of solution with changeovers

Thus we have

$$\sum_{i=1}^{m} y_1^i = 1$$

$$\sum_{j=1}^{m} \chi_t^{ij} = y_{t-1}^i \text{ for all } i, t$$

$$\sum_{i=1}^{m} \chi_t^{ij} = y_t^j \text{ for all } j, t$$

$$z_t^j + \chi_t^{jj} = y_t^j \text{ for all } j, t$$

$$y, z \in B^{mn}, \chi \in B^{m^2 n}$$

This formulation has an order of magnitude fewer constraints than (13.25)–(13.27), and it is as strong as possible. In what sense?

Heuristics. Relax-and-fix heuristics, in which, for example, variables in earlier periods are considered more important, provide one way of finding feasible solutions to large problems. Another approach is to consider the values of the set-up variables $\tilde{y}$ in the linear programming solution as probabilities, and thus to construct a solution in which, for each period t, the item i produced is found by selecting i randomly with probabilities $\tilde{y}_t^i$ for $i = 1, \ldots, m$.

13.6 A MULTIPLEXER ASSIGNMENT PROBLEM

We consider two variants of a problem that arose in the telecommunications industry.

Problem A. Given a graph $G = (V, E)$, demands $d_e > 0$ measuring the communications required between nodes i, j where $e = (i, j) \in E$, and an unlimited number of identical rings of capacity C, the demands must be assigned unsplit to the rings without exceeding capacity. If a demand d_e with $e = (i, j)$ is assigned to a certain ring, multiplexers costing α each must be installed at nodes i and j on that ring. The problem is to find a feasible assignment of demands to rings, so that the cost/number of multiplexers installed is minimized.

Problem B. Additional constraints are imposed on Problem A based on the structure of the multiplexers. Specifically a multiplexer contains eight slots, each of capacity $C_1 = 16$. By adding a special card to a slot, costing β, the capacity of a slot is increased to $C_2 = 63$. The problem is now to minimize the total cost of the multiplexers and the special cards.

We start by presenting a natural formulation for Problems A and B. Let $k = 1, \ldots, K$ denote the set of rings. We take as variables

$y_e^k = 1$ if demand d_e is assigned to ring k, and
$x_i^k = 1$ if a multiplexer is installed at node i of ring k,

and obtain as formulation A:

$$\min \alpha \sum_{k=1}^{K} \sum_{i \in V} x_i^k \qquad (13.33)$$
$$\sum_{k=1}^{K} y_e^k = 1 \text{ for } e \in E \qquad (13.34)$$
$$\sum_{e \in E} d_e y_e^k \leq C \text{ for } k = 1, \ldots, K \qquad (13.35)$$
$$y_e^k \leq x_i^k, y_e^k \leq x_j^k \text{ for } e = (i,j) \in E, k = 1, \cdots, K \qquad (13.36)$$
$$x_i^k \in B^1 \text{ for } i \in V, k = 1, \ldots, K \qquad (13.37)$$
$$y_e^k \in B^1 \text{ for } e \in E, k = 1, \ldots, K \qquad (13.38)$$

Here constraint (13.34) assigns each demand to some ring, (13.35) ensures that the ring capacity is not exceeded, and (13.36) that multiplexers are installed where required.

In Problem B, it is necessary to add additional variables

z_i^k the number of slots of capacity C_1 used at node i on ring k, and
w_i^k the number of slots of capacity C_2 used at node i on ring k.

The objective is now modified to become

$$\min \alpha \sum_{k=1}^{K} \sum_{i \in V} x_i^k + \beta \sum_{k=1}^{K} \sum_{i \in V} w_i^k,$$

and additional constraints

$$\sum_{e \in \delta(i)} d_e y_e^k \leq 16 z_i^k + 63 w_i^k \text{ for } i \in V, k = 1, \ldots, K$$
$$z_i^k + w_i^k \leq 8 x_i^k \text{ for } i \in V, k = 1, \ldots, K$$
$$z_i^k, w_i^k \in Z_+^1 \text{ for } i \in V, k = 1, \ldots, K$$

are needed to ensure that enough capacity is available at each node, and that the available number of slots is not exceeded.

For these two problems it is not clear a priori whether a branch-and-bound-based MIP system has any chance of producing reasonable solutions. Thus we first need to choose an algorithmic approach that will provide useful bounds, and then perhaps look further at the quality of the formulation. We again proceed via a series of questions. We start by examining Problem B.

Q1. Does Problem B have special structure?

A1. It is readily seen that problem B is of the form

$$\sum_{k=1}^{K} y_e^k = 1 \text{ for } e \in E$$

$$(y^k, x^k, z^k, w^k) \in X^k \text{ for } k = 1, \ldots, K$$

with each of the sets X^k identical.

Q2. What algorithmic approach does this suggest?

A2. From Chapters 10 and 11, we might consider a Lagrangian relaxation approach or a column generation approach, where the subproblem involves optimizing over X^k, or a cutting plane approach with separation over conv(X^k).

As we prefer where possible to use a general MIP system, we consider first the possibility of approximating conv(X^k) with valid inequalities.

Q3. Can we find valid inequalities for conv(X^k)?

A3. X^k contains 0–1 knapsack and integer knapsack constraints plus precedence constraints, but nothing else is apparent.

We also look at the complete problem.

Q4. Can we find global lower bounds on the number of multiplexers, or special slots required, either in total or at a specific node?

A4a. The total demand at node i is $D(i) = \sum_{e \in \delta(i)} d_e$. Assuming that only one multiplexer can be installed at each node on a ring, the maximum capacity of a multiplexer is $\min(C, 8 \times 63)$. Thus we obtain the valid inequality $\sum_{k=1}^{K} x_i^k \geq \lceil D(i)/\min(C, 504) \rceil$ for each node $i \in V$.

A4b. Working with slots in place of multiplexers, we have that

$$16 \sum_{k=1}^{K} z_i^k + 63 \sum_{k=1}^{K} w_i^k \geq D(i),$$

which after setting $Z = \sum_{k=1}^{K} z_i^k$ and $W = \sum_{k=1}^{K} w_i^k$ can be rewritten as the set $\{(Z,W) \in Z_+^2 : 16Z + 63W \geq D(i)\}$. Dividing by 16 we immediately obtain the Chvátal-Gomory inequality

$$Z + 4W \geq \lceil D(i)/16 \rceil.$$

It is also simple in practice to find all the inequalities required to represent the convex hull of a set in two variables.

Q5. What can be done about the symmetry, namely the fact that each ring is identical, and thus many feasible solutions are essentially the same?

A5. Constraints can be added to break the symmetry. For instance, one can impose that the multiplexer vector $(x_1^k, \ldots, x_{|V|}^k)$ for ring k be lexicographically as large as the corresponding vector for ring $k+1$. As only one multiplexer is allowed per node per ring, a sufficient set of constraints is

$$\begin{aligned}
x_1^k &\geq x_1^{k+1} \\
x_1^k + x_2^k &\geq x_1^{k+1} + x_2^{k+1} \\
2x_1^k + x_2^k + x_3^k &\geq 2x_1^{k+1} + x_2^{k+1} + x_3^{k+1} \\
4x_1^k + 2x_2^k + x_3^k + x_4^k &\geq 4x_1^{k+1} + 2x_2^{k+1} + x_3^{k+1} + x_4^{k+1} \\
&\cdots
\end{aligned}$$

Justify.

Q6. What branch-and-bound options should be considered?

A6. Given the relative importance of the variables, it appears natural to give the x_i^k variables highest priority for branching followed by the w_i^k variables. The reply to Q4 also suggests the idea of explicitly introducing aggregate variables $X^i = \sum_{k=1}^{K} x_i^k$ with W^i and Z^i defined similarly. In this case these variables should have higher priority than the corresponding individual variables.

Now we turn back to Problem A. As it has the same structure as Problem B, but no tightening of the formulation is apparent, decomposing as suggested in A2 appears a possibility.

Q7. Is Lagrangian relaxation or column generation a suitable algorithm for Problem A?

A7a. Criteria for evaluating the pros and cons of using Lagrangian relaxation were given in Section 10.5. Because the rings are identical, there is only one subproblem to be solved that takes the form

$$\min\{\alpha \sum_{i \in V} x_i - \sum_{e \in E} \pi_e y_e : (x,y) \in X\}$$

where $X = \{(x,y) \in B^{|V|} \times B^{|E|} : \sum_{e \in E} d_e y_e \leq C, y_e \leq x_i, y_e \leq x_j$ for $e = (i,j) \in E\}$.

We consider the difficulty of the subproblem, the difficulty of solving the Lagrangian dual, and the strength of the dual bound. Though the subproblem is $\mathcal{NP}$-hard, the graphs are typically small so the subproblem should be solved quickly by branch-and-bound. The only indication of the difficulty of convergence of the subgradient algorithm is the number of dual variables $|E|$. Finally as the constraints defining the set X certainly do not define $\text{conv}(X)$, the Lagrangian dual bound may well be significantly stronger than that provided by the linear programming relaxation of Formulation A. In practice this turns out to be the case, and the bound is very close to the optimal integer programming value.

A7b. Taking a column generation approach, the subproblem to be solved is the same as above, and the bound provided by the Linear Programming Master is the same as that of the Lagrangian dual. Implementing a column generation algorithm may require more work than implementing Lagrangian relaxation unless one of the codes specially designed for column generation is used. On the other hand the column generation procedure is more robust, and less likely to have difficulties of convergence. It may also provide good feasible solutions en route.

Q8. What about heuristics?

A8. Problem A is an edge-partitioning problem. The simulated annealing or tabu search heuristics proposed for node partitioning in Section 12.3 can be readily adapted to this problem as feasibility of a set of edges (demands) can be tested very easily. One possibility is to take as a solution an edge partition in which the demand on each ring is feasible, and as objective function the number of multiplexers required. As neighborhood one can take the set of feasible edge partitions obtained either by moving a single edge to another ring, or by interchanging a pair of edges lying on different rings.

13.7 NOTES

13.2 Details of many of the recent mathematical programming systems and modeling languages can be found in [ORMS97]. Four of the more well-known languages are AMPL, GAMS, LINGO and MP-MODEL(XPRESS), and three of the major MIP systems are CPLEX, OSL and XPRESS. The latter also have optimization subroutine libraries. There is now a movement toward the development of branch-and-cut sytems, and MINTO [SavNem93], MIPO [BalasCerCor96], and $bc - opt$ [Cordetal97] provide both 0-1 knapsack and mixed integer separation routines as part of a branch-and-cut solver for general IPs and MIPs.

For researchers, MINTO [SavNem93] and ABACUS [Thi95] are designed to facilitate the development of custom-built branch-and-cut or branch-and-price systems. EMOSL [EM97] is an extended modeling and optimization library permitting interaction between the modeling language and the optimizer.

13.3 A survey of cutting plane results for ULS and CLS can be found in [PocWol95].

13.4 The transformations in this section are all well-known. The multicommodity reformulation appears in [RarCho79] but was almost certainly known earlier. Its projection is analyzed in [RarWol93]. The dicut formulation for branchings is from [Edm67], and the dicut formulation for Steiner trees has been examined recently in [Goe94].

13.5 The model presented here is based on a variety of real applications. Recent results for models with start-ups can be found in [Con96]. The modeling of changeovers is from [KarmSch85]; see also [Wol97].

13.6 More details on the multiplexer assignment problem and its solution appear in [Sutetal98] and [Belvetal98].

Exercise 13.7 is a reduced version of a problem from a competition Whizzkids'96 that appeared in a Dutch newspaper [Whi96].

13.8 EXERCISES

1. Consider a unit commitment problem with 5 generators and 12 (2-hour) time periods. Period 1 follows on again from period 12, and the pattern is repeated daily. $d = (50, 60, 50, 100, 80, 70, 90, 60, 50, 120, 110, 70)$ are the demands per period, and the reserve is 1.2 times the demand, so the total capacity of the generators switched on in any period must be at least this reserve. The capacity of the generators is $C = (12, 12, 35, 50, 75)$, and their minimum levels of production are $L = (2, 2, 5, 20, 40)$. In addition each generator must stay on for at least two periods. The ramping constraints only apply to the fifth generator — when on in two successive periods, the output cannot increase by more than 20 from one period to the next, and cannot decrease by more than 15. The costs are approximate. The main cost is a start-up cost $g = (100, 100, 300, 400, 800)$. There are also fixed costs $f = (1, 1, 5, 10, 15)$ and variables costs $p = (10, 10, 4, 3, 2)$ in each period that a generator is on.

Formulate and solve with a mixed integer programming system. Try to minimize the number of nodes in the tree.

2. **Frequency Assignment.** Frequencies from the range $\{1,\ldots,6\}$ must be assigned to 10 stations. For each pair of stations, there is an interference parameter which is the minimum amount by which the frequencies of the two stations must differ. The pairs with nonzero parameter are given below:

$$
\begin{array}{cccccccccc}
e = & (1,2) & (2,3) & (3,4) & (4,5) & (5,6) & (6,7) & (7,8) & (8,9) \\
    & 1 & 2 & 4 & 2 & 3 & 1 & 1 & 2 \\
e = & (9,10) & (1,8) & (2,10) & (3,10) & (5,10) & (7,10) & (2,5) \\
    & 3 & 2 & 1 & 1 & 4 & 2 & 2
\end{array}
$$

The goal is to minimize the difference between the smallest and largest frequency assigned.

3. Find a minimum cost directed Steiner tree in the complete digraph on $n = 10$ nodes with root node $r = 1$, terminal nodes 2,5,7,8,10, and arc costs

$$
\begin{pmatrix}
12 & 6 & 7 & 4 & 11 & 8 & 4 & 5 & 7 & 15 \\
9 & 4 & 2 & 26 & 12 & 4 & 7 & 11 & 4 & 28 \\
6 & 3 & 8 & 2 & 12 & 15 & 19 & 3 & 7 & 9 \\
21 & 33 & 24 & 52 & 2 & 19 & 6 & 2 & 9 & 15 \\
6 & 3 & 5 & 8 & 4 & 3 & 2 & 7 & 4 & 12 \\
14 & 17 & 32 & 24 & 15 & 11 & 22 & 28 & 9 & 6 \\
8 & 3 & 5 & 2 & 3 & 4 & 7 & 8 & 10 & 3 \\
31 & 24 & 46 & 52 & 43 & 13 & 24 & 27 & 61 & 21 \\
2 & 3 & 4 & 3 & 5 & 7 & 9 & 3 & 5 & 9 \\
21 & 24 & 13 & 38 & 67 & 94 & 24 & 3 & 26 & 23
\end{pmatrix}
$$

4. Consider an instance of a constant capacity lot-sizing model with $n = 4$ periods, an initial stock of 8 units, demands $d = (2, 3, 6, 5)$, lower bounds on stocks of $(5, 2, 3, 0)$ units, and production capacity $C = 5$. Convert the instance to an equivalent one with no initial stock and no zero lower bounds on the stocks. Describe an algorithm for the general case.

5.* Show that the inequality

$$\sum_{j \in S} x_j \leq \sum_{j \in S}(\sum_{i=j}^{l} d_i)y_j + s_l$$

is facet-defining for X^{ULS} where $1 \leq l \leq n$ and $S \subseteq \{1,\ldots,l\}$. When is the more general inequality of Exercise 8.6 facet-defining?

6. Use the fact that any $s - t$ flow can be represented as a sum of flows on individual $s - t$ paths to reformulate the fixed charge network flow problem

of Section 13.4. Does this lead to any new algorithms for the problem?

7. Two cyclists must deliver newspapers in Manhattan. They pick them up at 6 a.m. and have to deliver to 60 customers as soon as possible. It is well known that the streets in Manhattan form a rectangular grid, so we can assume that the distance between two points (x_1, x_2) and (y_1, y_2) is $\mid x_1 - y_1 \mid + \mid x_2 - y_2 \mid$. The coordinates of the sixty customers are

$$\begin{array}{ccccc}
(17, 310) & (39, 85) & (48, 403) & (49, 444) & (55, 153) \\
(59, 250) & (59, 476) & (62, 353) & (81, 441) & (85, 367) \\
(85, 419) & (89, 418) & (105, 376) & (109, 258) & (110, 441) \\
(110, 447) & (118, 413) & (120, 49) & (120, 451) & (120, 459) \\
(122, 104) & (133, 410) & (142, 439) & (145, 412) & (146, 364) \\
(161, 190) & (161, 414) & (161, 434) & (162, 458) & (165, 374) \\
(167, 399) & (178, 409) & (179, 265) & (179, 365) & (179, 427) \\
(182, 359) & (184, 76) & (184, 198) & (185, 124) & (186, 169) \\
(186, 440) & (188, 63) & (194, 433) & (197, 352) & (200, 376) \\
(211, 462) & (212, 140) & (222, 181) & (223, 21) & (223, 328) \\
(233, 27) & (235, 405) & (239, 229) & (276, 231) & (284, 362) \\
(286, 24) & (292, 148) & (299, 188) & (302, 184) & (317, 237)
\end{array}$$

The depot is at $(375, 375)$. Assuming that both cyclists cover 300 distance units per hour, what is the earliest time by which every newspaper can be delivered?

8. A Multiplexer Assignment Instance. 5 nodes (communications centers) are linked on a network. Demands between nodes are

$$(d_e) = \begin{pmatrix} - & 86 & 45 & 65 & 116 \\ - & - & 9 & 0 & 41 \\ - & - & - & 104 & 126 \\ - & - & - & - & 8 \\ - & - & - & - & - \end{pmatrix}.$$

An unlimited number of fiber optic cables can be run through the five nodes in a ring. If part of the demand between nodes i and j is placed on some ring, add/drop multiplexers (ADMs) must be placed at nodes i and j. The total demand assigned to a ring (assuming 100% protection) cannot exceed 256. Each ADM has eight slots. The capacity of each slot is 16, unless a terminal multiplexer (TM) is installed in the slot, in which case its capacity is 63. The cost of an ADM is 3 and of a TM is 1. Find an assignment of the demands to the rings that minimizes the total cost.

9. The Traveling Salesman Problem with Time Windows. A truck driver must deliver to 9 customers on a given day, starting and finishing at the depot. Each customer $i = 1, \ldots, 9$ has a time window $[r_i, d_i]$ and an unloading time p_i. The driver must start unloading at client i during the specified

time interval. If she is early, she has to wait till time r_i before starting to unload. Node 0 denotes the depot, and c_{ij} the time to travel between nodes i and j for $i, j \in \{0, 1, \ldots, 9\}$. The data are $p = (0, 1, 5, 9, 2, 7, 5, 1, 5, 3)$, $r = (0, 2, 9, 4, 12, 0, 23, 9, 15, 10)$, $d = (150, 45, 42, 40, 150, 48, 96, 100, 127, 66)$, and

$$(c_{ij}) = \begin{pmatrix} - & 5 & 4 & 4 & 4 & 6 & 3 & 2 & 1 & 8 \\ 7 & - & 2 & 5 & 3 & 5 & 4 & 4 & 4 & 9 \\ 3 & 4 & - & 1 & 1 & 12 & 4 & 3 & 11 & 6 \\ 2 & 2 & 3 & - & 2 & 23 & 2 & 9 & 11 & 4 \\ 6 & 4 & 7 & 2 & - & 9 & 8 & 3 & 2 & 1 \\ 1 & 4 & 6 & 7 & 3 & - & 8 & 5 & 7 & 4 \\ 12 & 32 & 5 & 12 & 18 & 5 & - & 7 & 9 & 6 \\ 9 & 11 & 4 & 12 & 32 & 5 & 12 & - & 5 & 22 \\ 6 & 4 & 7 & 3 & 5 & 8 & 6 & 9 & - & 5 \\ 4 & 6 & 4 & 7 & 3 & 5 & 8 & 6 & 9 & - \end{pmatrix}.$$

10. **Lot-Sizing with Minimum Batch Sizes and Cleaning Times.** Consider the problem of production of a single item. Demands $\{d_t\}_{t=1}^n$ are known in advance. Production capacity in each time period is C, but in the last period of a production sequence it is reduced to $\tilde{C}$. In each production sequence, at least a minimum amount P must be produced, and production is at full capacity in all but the first and last periods of a production sequence. There is a fixed cost of f for each period in which the item is produced, the storage costs are h per item per period, and the backlogging costs g. Solve an instance with $n = 12$, $d = (5, 4, 0, 0, 6, 3, 2, 0, 0, 4, 9, 0)$, $C = 7$, $\tilde{C} = 4$, $f = 50$, $h = 1$, $g = 10$, and $P = 19$.

References

AarLen97. E.H.L. Aarts and J.K. Lenstra, eds., *Local Search in Combinatorial Optimization*, Wiley, Chichester (1997).

AarVer97. E.H.L. Aarts and M. Verhoeven, Local Search, Ch. 11 in [AarLen97].

AghMagWol95. E.H Aghezzaf, T.L. Magnanti and L.A. Wolsey, Optimizing Constrained Subtrees of Trees, *Mathematical Programming* **71**, 113–126 (1995).

AhuMagOrl93. R. K. Ahuja, T.L. Magnanti and J.B. Orlin, *Network Flows*, Prentice Hall, Englewood Cliffs, NJ, (1993).

Appetal95. D. Applegate, W.J. Cook, R. Bixby and V. Chvátal, Finding Cuts in the TSP, DIMACS Technical Report 95-05, Rutgers University, New Brunswick, NJ, March 1995.

Balaketal95. A. Balakrishnan, T.L. Magnanti and R.T. Wong, A Decomposition Algorithm for Local Access Telecommunication Network Expansion Planning, *Operations Research* **43**, 58–66 (1995).

Balas65. E. Balas, An Additive Algorithm for Solving Linear Programs with 0-1 Variables, *Operations Research* **13**, 517–546 (1965).

Balas75a. E. Balas, Disjunctive Programs: Cutting Planes from Logical Conditions, in O.L. Mangasarian et al., eds., *Nonlinear Programing, Vol. 2*, Academic Press, New York, pp. 279–312 (1975).

Balas75b. E. Balas, Facets of the Knapsack Polytope, *Mathematical Programming* **8**, 146–164 (1975).

Balas89. E. Balas, The Prize Collecting Traveling Salesman Problem, *Networks* **19**, 621–636 (1989).

Balas95. E. Balas, New Classes of Efficiently Solvable Generalized Traveling Salesman Problems, MSSR-611, GSIA, Carnegie-Mellon University, Pittsburgh, March 1995.

BalasCerCor93. E. Balas, S. Ceria and G. Cornuéjols, A Lift-and-Project Cutting Plane Algorithm for Mixed 0-1 Programs, *Mathematical Programming* **58**, 295–324 (1993).

BalasCerCor96. E. Balas, S. Ceria and G. Cornuéjols, Mixed 0–1 Programming by Lift-and-Project in a Branch-and-Cut Framework, *Management Science* **42**, 1229–1246 (1996).

Balasetal96. E. Balas, S. Ceria, G. Cornuéjols and G. Natraj, Gomory Cuts Revisited, *Operations Research Letters* **19**, 1–9 (1996).

Balletal95a. M.O. Ball, T.L. Magnanti, C.L. Monma and G.L. Nemhauser, eds., Handbooks in Operations Research and Management Science, Vol. 7, *Network Models*, North-Holland, Amsterdam (1995).

Balletal95b. M.O. Ball, T.L. Magnanti, C.L. Monma and G.L. Nemhauser, eds., Handbooks in Operations Research and Management Science, Vol. 8, *Network Routing*, North-Holland, Amsterdam (1995).

BarEdmWol86. I. Barány, J. Edmonds and L.A. Wolsey, Packing and Covering a Tree by Subtrees, *Combinatorica* **6**, 245–257 (1986).

Barnetal94. C. Barnhart, E.L. Johnson, G.L. Nemhauser and M.W.P. Savelsbergh, Branch and Price: Column Generation for Solving Huge Integer Programs, Computational Optimization Center COC-94-03, Georgia Institute of Technology, Atlanta, February 1994 (revised May 1995).

Bau97. P. Bauer, The Circuit Polytope: Facets, *Mathematics of Operations Research* **22** 110–145 (1997).

Beal79. E.M.L. Beale, Branch and Bound Methods for Mathematical Programming Systems, *Annals of Discrete Mathematics* **5**, 201–219 (1979).

BealTom70. E.M.L. Beale and J.A. Tomlin, Special Facilities in a General Mathematical Programming System for Nonconvex Problems Using Ordered Sets of Variables, in J. Lawrence ed., *Proceedings of the Fifth Annual Conference on Operational Research*, Tavistock Publications, pp. 447–454 (1970).

Beas93. J.E. Beasley, Lagrangean Relaxation, Ch. 6 in *Modern Heuristic Techniques for Combinatorial Problems*, Blackwell Scientific Publications, Oxford (1993).

Beas96. J.E. Beasley, ed., *Advances in Linear and Integer Programming*, Oxford University Press, Oxford (1996).

Beas97. J.E. Beasley, A Genetic Algorithm for the Generalised Assignment Problem, *Computers and Operations Research* **24**, 17–23 (1997).

Bell57. R.E. Bellman, *Dynamic Programming*, Princeton University Press, Princeton (1957).

Belvetal98. G. Belvaux, N. Boissin, A. Sutter and L.A. Wolsey, Optimal Placement of Add/Drop Multiplexers: Static and Dynamic Models, *European Journal of Operational Research* (to appear) (1998).

Ben62. J.F. Benders, Partitioning Procedures for Solving Mixed Variables Programming Problems, *Numerische Mathematik* **4**, 238–252 (1962).

Ber57. C. Berge, Two Theorems in Graph Theory, *Proceedings of the National Academy of Science* **43**, 842–844 (1957).

Ber73. C. Berge, *Graphs and Hypergraphs*, North-Holland, Amsterdam (1973).

BieGun98. D. Bienstock and O. Günlük, Capacitated Network Design — Polyhedral Structure and Computation, *ORSA Journal on Computing* (to appear) (1998).

BonMur76. J.A. Bondy and U.S.R. Murty, *Graph Theory with Applications*, Macmillan, London (1976).

BreMitWil73. A.L. Brearley, G. Mitra and H.P. Williams, An Analysis of Mathematical Programs Prior to Applying the Simplex Method, *Mathematical Programming* **7**, 263–282 (1973).

CapFis97. A. Caprara and M. Fischetti, Branch and Cut Algorithms, Ch. 4 in [DelAMafMar97].

Cerietal95. S. Ceria, C. Cordier, H. Marchand and L.A. Wolsey, Cutting Planes for Integer Programs with General Integer Variables, *Mathematical Programming* (to appear) (1998).

Cern85. V. Cerny, A Thermodynamical Approach to the Travelling Salesman Problem: An Efficient Simulation Algorithm, *Journal of Optimization Theory and Applications* **45**, 41–51 (1985).

Chr76. N. Christofides, Worst Case Analysis of a New Heuristic for the Travelling Salesman Problem, Report 388, GSIA, Carnegie-Mellon University, (1976).

ChrMinTot81. N. Christofides, A. Mingozzi and P. Toth, State Space Relaxation Procedures for the Computation of Bounds to Routing Problems, *Networks* **11**, 145–164 (1981).

Chv73. V. Chvátal, Edmonds Polytopes and a Hierarchy of Combinatorial Problems, *Discrete Mathematics* **4**, 305–337 (1973).

Chv83. V. Chvátal, *Linear Programming*, Freeman, New York (1983).

CloNad93. J.M. Clochard and D. Naddef, Using Path Inequalities in a Branch and Cut Code for the Symmetric Travelling Salesman Problem, in G. Rinaldi and L.A., Wolsey eds., *Proceedings 3rd IPCO Conference*, Louvain-la-Neuve, pp. 291–311 (1993).

Con96. M. Constantino, A Cutting Plane Approach to Capacitated Lot-Sizing with Start-up Costs, *Mathematical Programming* **75**, 353–376 (1996).

Con98. M. Constantino, Lower Bounds in Lot-Sizing Models: a Polyhedral Study, *Mathematics of Operations Research* **23**, 101–118 (1998).

Coo71. S.A. Cook, The Complexity of Theorem-Proving Procedures, *Proceedings of the 3rd Annual ACM Symposium on Theory of Computing Machinery*, ACM, 151–158 (1971).

CooCunetal97. W.J. Cook, W.H. Cunningham, W.R. Pulleyblank and A. Schrijver, *Combinatorial Optimization*, Wiley, New York (1997).

CooKanSch90. W.J. Cook, R. Kannan and A. Schrijver, Chvátal Closures for Mixed Integer Programming Problems, *Mathematical Programming* **47**, 155–174 (1990).

CooLovSey95. W.J. Cook, L. Lovász and P. Seymour, eds., *Combinatorial Optimization*, DIMACS Series in Discrete Mathematics and Computer Science, AMS (1995).

CooRutetal93. W. Cook, T. Rutherford, H.E. Scarf and D. Shallcross, An Implementation of the Generalized Basis Reduction Algorithm for Integer Programming, *ORSA Journal of Computing* **5**, 206–212(1993).

Cordetal97. C. Cordier, H. Marchand, R. Laundy and L.A. Wolsey, $bc - opt$: A Branch-and-Cut Code for Mixed Integer Programs, CORE Discussion Paper 9778, Université Catholique de Louvain, October 1997.

CorNem78. G. Cornuéjols and G.L. Nemhauser, Tight Bounds for Christofides' Travelling Salesman Heuristic, *Mathematical Programming* **14**, 116–121 (1978).

CorSriThi91. G. Cornuéjols, R. Sridharan and J.M. Thizy, A Comparison of Heuristics and Relaxations for the Capacitated Plant Location Problem, *European Journal of Operational Research* **50**, 280–297 (1991).

REFERENCES 249

CreKan95. P. Crescenzi and V. Kann, A Compendium of $\mathcal{NP}$ Optimization Problems, Technical Report SI/RR-95/02, Dipartimento di Scienze dell'Informazione, University of Rome, "La Sapienza" (1995).

Cro58. G.A. Croes, A Method for Solving Travelling Salesman Problems, *Operations Research* **6**, 791–812 (1958).

CroJohPad83. H. Crowder, E.L. Johnson and M.W. Padberg, Solving Large Scale Zero-One Linear Programming Problems, *Operations Research* **31**, 803–834 (1983).

Dak65. R.J. Dakin, A Tree Search Algorithm for Mixed Integer Programming Problems, *Computer Journal* **8**, 250–255 (1965).

Dan57. G.B. Dantzig, Discrete Variable Extremum Problems, *Operations Research* **5**, 266–277 (1957).

Dan63. G.B. Dantzig, *Linear Programming and Extensions*, Princeton University Press, Princeton (1963).

DanForFul56. G.B. Dantzig, L.R. Ford and D.R. Fulkerson, A Primal-Dual Algorithm for Linear Programming, in H. Kuhn and A.W. Tucker, eds., *Linear Inequalities and Related Systems*, Princeton University Press, Princeton (1956).

DanFulJoh54. G.B. Dantzig, D.R. Fulkerson and S.M. Johnson, Solution of a Large Scale Traveling Salesman Problem, *Operations Research* **2**, 393–410 (1954).

DanWol60. G.B. Dantzig and P. Wolfe, Decomposition Principle for Linear Programs, *Operations Research* **8**, 101–111 (1960).

DelAMafMar97. M. Dell'Amico, F. Maffioli and S. Martello, eds., *Annotated Bibliographies in Combinatorial Optimization*, Wiley, Chichester (1997).

Den82. E.V. Denardo, *Dynamic Programming Models and Applications* Prentice-Hall (1982).

Desetal95. J. Desrosiers, Y. Dumas, M.M. Solomon and F. Soumis, Time Constrained Routing and Scheduling, Ch. 2 in [Balletal95b].

DesSou89. J. Desrosiers and F. Soumis, A Column Generation Approach to the Urban Transit Crew Scheduling Problem, *Transportation Science* **23**, 1–13 (1989).

DesSouDes84. J. Desrosiers, F. Soumis and M. Desrochers, Routing with Time Windows by Column Generation, *Networks* **14**, 545–565 (1984).

Edm65a. J. Edmonds, Paths, Trees and Flowers, *Canadian Journal of Mathematics* **17**, 449–467 (1965).

Edm65b. J. Edmonds, Maximum Matching and a Polyhedron with 0-1 Vertices, *Journal of Research of the National Bureau of Standards* **69B**, 125–130 (1965).

Edm67. J. Edmonds, Optimum Branchings, *Journal of Research of the National Bureau of Standards* **71B**, 233–240 (1967).

Edm70. J. Edmonds, Submodular Functions, Matroids and Certain Polyhedra, in R. Guy, ed., *Combinatorial Structures and Their Applications, Proceedings of the Calgary International Conference*, Gordon and Breach, pp. 69–87 (1970).

Edm71. J. Edmonds, Matroids and the Greedy Algorithm, *Mathematical Programming* **1**, 127–136 (1971).

EdmGil77. J. Edmonds and R. Giles, A Min-Max Relation for Submodular Functions on Graphs, *Annals of Discrete Mathematics* **1**, 185–204 (1997).

EM97. XPRESS-MP Extended Modeling and Optimisation Subroutine Library, Reference Manual, Release 10, Dash Associates, Blisworth House, Blisworth, Northants (1997).

Erl78. D. Erlenkotter, A Dual-Based Procedure for Uncapacitated Facility Location, *Operations Research* **26**, 992–1009 (1978).

Eve63. H. Everett III, Generalized Lagrange Multiplier Method for Solving Problems of Optimal Allocation of Resources, *Operations Research* **11**, 399–417 (1963).

Fis81. M.L. Fisher, The Lagrangean Relaxation Method for Solving Integer Programming Problems, *Management Science* **27**, 1–18 (1981).

FisRin88. M.L. Fisher and A.H.G. Rinnooy Kan, eds., The Design and Analysis of Heuristics, *Management Science* **34**, 263–429 (1988).

Fle90. D. Fleischmann, The Discrete Lot-Sizing and Scheduling Problem, *European Journal of Operational Research* **44**, 337–348 (1990).

FloKle71. M. Florian and M. Klein, Deterministic Production Planning with Concave Costs and Capacity Constraints, *Management Science* **18**, 12–20 (1971).

ForFul56. L.R. Ford, Jr. and D.R. Fulkerson, Maximum Flow through a Network, *Canadian Journal of Mathematics* **8**, 399–404 (1956).

ForFul62. L.R. Ford, Jr. and D.R. Fulkerson, *Flows in Networks*, Princeton University Press, Princeton (1962).

FulGro65. D.R. Fulkerson and D.A. Gross, Incidence Matrices and Interval Graphs, *Pacific Journal of Mathematics* **15**, 833–835 (1965).

GarJoh79. M.R. Garey and D.S. Johnson, *Computers and Intractability: A Guide to the Theory of NP-Completeness*, Freeman, San Francisco (1979).

Geo74. A.M. Geoffrion, Lagrangean Relaxation for Integer Programming, *Mathematical Programming Study* **2**, 82–114 (1974).

GeoMar72. A.M. Geoffrion and R.E. Marsten, Integer Programming Algorithms: A Framework and State-of-the Art Survey, *Management Science* **18**, 465–491 (1972).

Gho62. A. Ghouila-Houri, Caracterisation des Matrices Totalement Unimodulaires, *C.R. Academy of Sciences of Paris* **254**, 1192–1194 (1962).

GilGom61. P.C. Gilmore and R.E. Gomory, A Linear Programming Approach to the Cutting Stock Problem, *Operations Research* **9**, 849–859 (1961).

GilGom63. P.C. Gilmore and R.E. Gomory, A Linear Programming Approach to the Cutting Stock Problem: Part II, *Operations Research* **11**, 863–888 (1963).

GilGom66. P.C. Gilmore and R.E. Gomory, The Theory and Computation of Knapsack Functions, *Operations Research* **14**, 1045–1074 (1966).

Glo68. F. Glover, Surrogate Constraints, *Operations Research* **16**, 741–749 (1968).

Glo86. F. Glover, Future Paths for Integer Programming and Links to Artificial Intelligence, *Computers and Operations Research* **13**, 533–549 (1986).

Glo89. F. Glover, Tabu Search: Part I, *ORSA Journal on Computing* **1**, 190–206 (1989).

Glo90. F. Glover, Tabu Search: Part II, *ORSA Journal on Computing* **2**, 4–32 (1990).

Goe94. M.X. Goemans, The Steiner Polytope and Related Polyhedra, *Mathematical Programming* **63**, 157–182 (1994).

Gof77. J-L. Goffin, On the Convergence Rates of Subgradient Optimization Methods, *Mathematical Programming* **13**, 329–347 (1977).

Gol89. D.E. Goldberg, *Genetic Algorithms in Search, Optimization, and Machine Learning*, Addison-Wesley, Reading, MA (1989).

Gom58. R.E. Gomory, Outline of an Algorithm for Integer Solutions to Linear Programs, *Bulletin of the American Mathematical Society* **64**, 275–278 (1958).

Gom60. R.E. Gomory, An Algorithm for the Mixed Integer Problem, RM-2597, The Rand Corporation (1960).

Gom63. R.E. Gomory, An Algorithm for Integer Solutions to Linear Programs, in R. Graves and P. Wolfe, eds., *Recent Advances in Mathematical Programming*, McGraw-Hill, New York, pp. 269-302 (1963).

Gom65. R.E. Gomory, On the Relation between Integer and Non-Integer Solutions to Linear Programs, *Proceedings of the National Academy of Sciences* **53**, 260–265 (1965).

Gom69. R.E. Gomory, Some Polyhedra related to Corner Problems, *Linear Algebra and its Applications* **2**, 451–588 (1969).

GomHu61. R.E. Gomory and T.C. Hu, Multi-Terminal Network Flows, *SIAM Journal* **9**, 551–570 (1961).

GomJoh72. R.E. Gomory and E.L. Johnson, Some Continuous Functions related to Corner Polyhedra, *Mathematical Programming* **3**, 23–85 (1972).

Gra66. R.L. Graham, Bounds for Certain Multiprocessing Anomalies, *Bell Systems Technical Journal* **45**, 1563–1581 (1966).

GraRinZip93. S.C. Graves, A.H.G. Rinnooy Kan and P.H. Zipkin, eds., Handbooks in Operations Research and Management Science, Vol. 4, *Logistics of Production and Inventory*, North-Holland, Amsterdam (1993).

GroLie81. H. Gröflin and T. Liebling, Connected and Alternating Vectors: Polyhedra and Algorithms, *Mathematical Programming* **20**, 233–244 (1981).

GroLovSch81. M. Grötschel, L. Lovász and A. Schrijver, The Ellipsoid Method and its Consequences in Combinatorial Optimization, *Combinatorica* **1**, 169–197 (1981).

GroLovSch84. M. Grötschel, L. Lovász and A. Schrijver, Corrigendum to our Paper "The Ellipsoid Method and its Consequences in Combinatorial Optimization," *Combinatorica* **4**, 291–295 (1984).

GroLovSch88. M. Grötschel, L. Lovasz and A. Schrijver, *Geometric Algorithms and Combinatorial Optimization*, Springer, Berlin (1988).

GroPad75. M. Grötschel and M. W. Padberg, Partial Linear Characterizations of the Asymmetric Travelling Salesman Polytope, *Mathematical Programming* **8**, 378–381 (1975).

Gru67. B. Grunbaum, *Convex Polytopes*, Wiley, New York (1967).

Guetal95. Z. Gu, G.L. Nemhauser and M.W.P. Savelsbergh, Lifted Cover Inequalities for 0–1 Integer Programs, LEC-96-05, Georgia Institute of Technology (1995).

Guetal96. Z. Gu, G.L. Nemhauser and M.W.P. Savelsbergh, Lifted Flow Cover Inequalities for Mixed 0–1 Integer Programs, LEC-96-05, Georgia Institute of Technology (1996).

GuiKim87. M. Guignard and S. Kim, Lagrangean Decomposition for Integer Programming: Theory and Applications, *RAIRO* **21**, 307–323 (1987).

GuiSpi81. M. Guignard and K. Spielberg, Logical Reduction Methods in Zero-One Programming, *Operations Research* **29**, 49–74 (1981).

GunPoc98. O. Günlük and Y. Pochet, Mixing Mixed-Integer Inequalities, CORE DP9811, Université Catholique de Louvain (1998).

HamJohPel75. P.L. Hammer, E.L. Johnson and U.N. Peled, Facets of Regular 0-1 Polytopes, *Mathematical Programming* **8**, 179–206 (1975).

HelKar62. M. Held and R.M. Karp: A Dynamic Programming Approach to Sequencing Problems. *Journal of SIAM* **10**, 196–210 (1962).

HelKar70. M. Held and R.M. Karp, The Traveling Salesman Problem and Minimum Spanning Trees, *Operations Research* **18**, 1138–1162 (1970).

HelKar71. M. Held and R.M. Karp, The Traveling Salesman Problem and Minimum Spanning Trees: Part II, *Mathematical Programming* **1**, 6–25 (1971).

HelWolCro74. M. Held, P. Wolfe and H.P. Crowder, Validation of Subgradient Optimization, *Mathematical Programming* **6**, 62–88 (1974).

Hoc95. D.S. Hochbaum, ed., *Approximation Algorithms for $\mathcal{NP}$-hard Problems*, PWS Publishing, Boston (1995).

HofKru56. A.J. Hoffman and J.B. Kruskal, Integral Boundary Points of Convex Polyhedra, in H.W. Kuhn and A.W. Tucker, eds., *Linear Inequalities and Related Systems*, Princeton University Press, Princeton, pp. 223–246 (1956).

HofPad91. K. Hoffman and M. Padberg, Improving Representation of Zero-One Linear Programs for Branch-and-Cut, *ORSA Journal of Computing* **3**, 121–134 (1991).

Hol75. J.H. Holland, *Adaptation in Natural and Artificial Systems*, University of Michigan Press, Ann Arbor, MI, (1975).

IbaKim75. O.H. Ibarra and C.E. Kim, Fast Approximations Algorithms for the Knapsack and Sum of Subset Problems, *Journal of the ACM* **22**, 463–468 (1975).

Jer72. R.G. Jeroslow, Cutting Plane Theory: Disjunctive Methods, *Annals of Discrete Mathematics* **1**, 293–330 (1972).

Johetal74. D.S. Johnson, A. Demers, J.D. Ullman, M.R. Garey and R.L. Graham, Worst Case Performance Bounds for Simple One-Dimensional Packing Algorithms, *SIAM Journal on Computing* **3**, 299–325 (1974).

Joh80. E.L. Johnson, *Integer Programming — Facets, Subadditivity and Duality for Group and Semigroup Problems*, SIAM Publications, Philadelphia (1980).

JohMehNem93. E.L. Johnson, A. Mehrotra and G.L. Nemhauser, Min-Cut Clustering, *Mathematical Programming* **62**, 133–152 (1993).

JorNas86. K. Jornsten and M. Nasberg, A New Lagrangian Relaxation Approach to the Generalized Assignment Problem, *European Journal of Operational Research* **27**, 313–323 (1986).

JueReiRin95. M. Jünger, G. Reinelt and G. Rinaldi, The Travelling Salesman Problem, Ch. 4, 225–330 in [Balletal95a].

JueReiThi95. M. Jünger, G. Reinelt and S. Thienel, Practical Problem Solving with Cutting Plane Algorithms in Combinatorial Optimization, 11–152 in [CooLovSey95].

KarmSch85. U.S. Karmarkar and L.Schrage, The Deterministic Dynamic Product Cycling Problem, *Operations Research* **33**, 326–345 (1985).

Karp72. R.M. Karp, Reducibility among Combinatorial Problems, in R.E. Miller and J.W. Thatcher, eds., *Complexity of Computer Computations*, Plenum Press, New York, 85–103 (1972).

Karp75. R.M. Karp, On the Complexity of Combinatorial Problems, *Networks* **5**, 45–68 (1975).

Kiretal83. S. Kirkpatrick, C.D. Gelatt, Jr., and M.P. Vecchi, Optimization by Simulated Annealing, *Science* **220**, 671–680 (1983).

Kru56. J. B. Kruskal, On the Shortest Spanning Subtree of a Graph and the Traveling Salesman Problem, *Proceedings of the American Mathematical Society* **7**, 48–50 (1956).

KueHam63. A.A. Kuehn and M.J. Hamburger, A Heuristic Program for Locating Warehouses, *Management Science* **9**, 643–666 (1963).

Kuh55. H.W. Kuhn, The Hungarian Method for the Assignment Problem, *Naval Research Logistics Quarterly* **2**, 83–97 (1955).

LanDoi60. A.H. Land and A.G. Doig, An Automatic Method for Solving Discrete Programming Problems, *Econometrica* **28**, 497–520 (1960).

Law76. E.L. Lawler, *Combinatorial Optimizaton: Networks and Matroids*, Holt, Rinehart and Winston, New York (1976).

Lawetal85. E.L. Lawler, J.K. Lenstra, A.H.G. Rinnooy Kan and D.B. Shmoys, eds., *The Traveling Salesman Problem: A Guided Tour of Combinatorial Optimization*, Wiley, Chichester (1985).

Lemetal95. C. Lemaréchal, F. Pellegrino, A. Renaus and C. Sagastizabal, Bundle Methods Applied to the Unit Commitment Problem, Communication at the *17th IFIP TC7 Conference*, Prague, July 1995.

Len83. H.W. Lenstra Jr., Integer Programming with a Fixed Number of Variables, *Mathematics of Operations Research* **8**, 538–547 (1983).

LinSav97. J. Linderoth and M.W.P. Savelsbergh, A Computational Study of Search Strategies for Mixed Integer Programming, Report LEC 97-12, Georgia Institute of Technology (1997).

Litetal63. J.D.C. Little, K.G. Murty, D.W. Sweeney and C. Karel, An Algorithm for the Traveling Salesman Problem, *Operations Research* **11**, 972–989 (1963).

LovPlu86. L. Lovász and M.D. Plummer, *Matching Theory*, North-Holland, Amsterdam (1986).

LovSch91. L. Lovász and A. Schrijver, Cones of Matrices and Set Functions and 0-1 Optimization, *SIAM Journal on Optimization* **1**, 166–190 (1991).

LucBea96. A. Lucena and J.E. Beasley, Branch and Cut Algorithms, Ch. 5 in [Beas96].

MagWol95. T.L. Magnanti and L.A. Wolsey, Optimal Trees, in [Balletal95a].

MarTot90. S. Martello and P. Toth, *Knapsack Problems: Algorithms and Computer Implementations*, Wiley, Chichester (1990).

MarWol97. H. Marchand and L.A. Wolsey, The 0-1 Knapsack Problem with a Single Continuous Variable, CORE DP 9720, Louvain-la-Neuve, March 1997.

Metetal53. N. Metropolis, A. Rosenbluth, H. Rosenbluth, A. Teller and E. Teller, Equations of State Calculations by Fast Computing Machines, *Journal of Chemical Physics* **21**, 1087–1091 (1953).

MirFra90. P.B. Mirchandani and R.L. Francis, eds., *Discrete Location Theory*, Wiley, New York (1990).

Mit96. J.E. Mitchell, Interior Point Algorithms for Integer Programming, Ch. 6 in [Beas96].

MitTod92. J.E. Mitchell and M.J. Todd, Solving Combinatorial Optimization Problems Using Karmarkar's Algorithm, *Mathematical Programming* **56**, 245–285 (1992).

NemRinTod89. G.L. Nemhauser, A.H.G. Rinnooy Kan and M.J. Todd, eds., Handbooks in Operations Research and Management Science, Vol. 1, *Optimization*, North-Holland, Amsterdam (1989).

NemSavSig94. G.L. Nemhauser, M.W.P. Savelsbergh and G. Sigismondi, MINTO, a Mixed Integer Optimizer, *Operations Research Letters* **15**, 47–58 (1994).

NemTro74. G.L. Nemhauser and L.E. Trotter, Properties of Vertex Packing and Independence System Polyhedra, *Mathematical Programming* **6**, 48–61 (1974).

NemWol88. G.L. Nemhauser and L.A. Wolsey, *Integer and Combinatorial Optimization*, Wiley (1988).

NemWol89. G.L. Nemhauser and L.A. Wolsey, Integer Programming, Ch. 6 in [NemRinTod89]

NemWol90. G.L. Nemhauser and L.A. Wolsey, A Recursive Procedure for Generating All Cuts for 0-1 Mixed Integer Programs, *Mathematical Programming* **46**, 379–390 (1990).

NorRab59. R.Z. Norman and M.D. Rabin, An Algorithm for the Minimum Cover of a Graph, *Proceedings of the American Mathematical Society* **10**, 315–319 (1959).

OheLenRin85. M. O'hEigeartaigh, J.K. Lenstra and A.H.G. Rinnooy Kan, eds., *Combinatorial Optimization: Annotated Bibliographies*, Wiley, Chichester (1985).

ORMS97. Linear Programming Software Survey, *OR/MS Today* 56–63, April 1997

OsmLap96. I.H. Osman and G. Laporte, Metaheuristics: a Bibliography, *Annals of Operations Research* **63**, 513–623 (1996).

Pad73. M.W. Padberg, On the Facial Structure of Set Packing Polyhedra, *Mathematical Programming* **5**, 199–215 (1973).

PadRin91. M.W. Padberg and G. Rinaldi, A Branch and Cut Algorithm for Resolution of Large Scale Symmetric Traveling Salesman Problems, *SIAM Review* **33**, 60–100 (1991).

PadVanRWol85. M.W. Padberg, T.J. Van Roy and L.A. Wolsey, Valid Linear Inequalities for Fixed Charge Problems, *Operations Research* **33**, 842–861 (1985).

Pap94. C.H. Papadimitriou, *Computational Complexity*, Addison-Wesley, Reading, MA (1994).

PapSti82. C.H. Papadimitriou and K. Stieglitz, *Combinatorial Optimization: Algorithms and Complexity*, Prentice-Hall, Englewood Cliffs, NJ (1982).

PapYan84. C.H. Papadimitriou and M. Yannakakis, The Complexity of Facets (and Some Facets of Complexity), *Journal of Computing and System Science* **28**, 244–259 (1984).

ParRar88. G. Parker and R. Rardin, *Discrete Optimization*, Academic Press, New York (1988).

Pir96. M. Pirlot, General Local Search Methods, *European Journal of Operational Research* **92**, 493–511 (1996).

PocWol95. Y. Pochet and L.A. Wolsey, Algorithms and Reformulations for Lot-Sizing Problems, 245–294 in [CooLovSey95].

PreSha85. F.P. Preparata and M.I. Shamos, *Computational Geometry: An Introduction*, Springer, New York (1985).

Prim57. R.C. Prim, Shortest Connection Networks and Some Generalizations, *Bell System Technological Journal* **36**, 1389–1401 (1957).

Psa80. H.N. Psaraftis, A Dynamic Programming Approach for Sequencing Groups of Identical Jobs, *Operations Research* **28**, 1347–1359 (1980).

Pul83. W.R. Pulleyblank, Polyhedral Combinatorics, in *Mathematical Programming: The State of the Art*, A. Bachem, M. Grötschel and B. Korte, eds., Springer, Berlin, pp. 312–345 (1983).

Pul89. W.R. Pulleyblank, Polyhedral Combinatorics, in [NemRinTod89].

QueSch94. M. Queyranne and A.S. Schulz, Polyhedral Approaches to Machine Scheduling, Preprint 408/1994, Department of Mathematics, Technical University of Berlin, Berlin (1994).

RarCho79. R. Rardin and U. Choe, Tighter Relaxations of Fixed Charge Network Flow Problems, Industrial and Systems Engineering Report J-79-18, Georgia Institute of Technology (1979).

RarWol93. R. Rardin and L.A. Wolsey, Valid Inequalities and Projecting the Multicommodity Extended Formulation for Uncapacititated Fixed Charge Network Flow Problems, *European Journal of Operational Research* (1993).

Ree93. C.R. Reeves, ed., *Modern Heuristic Techniques for Combinatorial Problems*, Blackwell Scientific Publications, Oxford (1993).

ReiShe65. S. Reiter and G. Sherman, Discrete Optimizing, *J. Society of Industrial and Apllied Mathematics* **13**, 864–889 (1965).

REFERENCES

Rhy70. J.M.W. Rhys, A Selection Problem of Shared Fixed Costs and Network Flows, *Management Science* **17**, 200–207 (1970).

RooTerVia97. C. Roos, T. Terlaky and J-Ph. Vial, *Theory and Algorithms for Linear Optimization: An Interior Point Approach*, Wiley (1997).

RyaFos81. D.M. Ryan and B.A. Foster, An Integer Programming Approach to Scheduling, in A. Wren, ed., *Computer Scheduling of Public Transport Urban Passenger Vehicle and Crew Scheduling*, North-Holland, Amsterdam, 269–280 (1981).

SalMat89. H.M. Salkin and K. Mathur, *Foundations of Integer Programming*, North-Holland, Amsterdam (1989).

Sav93. M.W.P. Savelsbergh, A Branch and Price Algorithm for the Generalized Assignment Problem, Computational Optimization Center COC-93-02, Georgia Institute of Technology, Atlanta (1993).

Sav94. M.W.P. Savelsbergh, Preprocessing and Probing Techniques for Mixed Integer Programming Problems, *ORSA Journal of Computing* **6**, 445–454 (1994).

SavNem93. M.W.P. Savelsbergh and G.L. Nemhauser, Functional Description of MINTO, a Mixed INTeger Optimizer, Report COC-91-03A, Georgia Institute of Technology, Atlanta, Georgia (1993).

Sch80. A. Schrijver, On Cutting Planes, *Annals of Discrete Mathematics* **9**, 291–296 (1980).

Sch86. A. Schrijver, *Theory of Linear and Integer Programming*, Wiley, Chichester (1986).

Sey80. P.D. Seymour, Decomposition of Regular Matroids, *Journal of Combinatorial Theory* **B28**, 305–359 (1980).

SheAda90. H. Sherali and W. Adams, A Hierarchy of Relaxations between the Continuous and Convex Hull Representations for Zero-One Programming Problems, *SIAM Journal on Discrete Mathematics* **3**, 411–430 (1990).

Shm95. D.B. Shmoys, Computing Near-Optimal Solutions to Combinatorial Optimization Problems, 355–398 in [CooLovSey95].

Sie96. G. Sierksma, *Integer and Linear Programming: Theory and Practice*, Marcel Dekker (1996).

Spi69. K. Spielberg, Plant Location with Generalized Search Origin, *Management Science* **16**, 165–178 (1969).

Sta92. J. Stallaert, On the Polyhedral Structure of Capacitated Fixed Charge Network Problems, Ph.D. Thesis, Univ. of California (1992).

StoWit70. J. Stoer and C. Witzgall, Convexity and Optimization in Finite Dimensions, Springer (1970).

Sutetal98. A. Sutter, F. Vanderbeck and L.A. Wolsey, Optimal Placement of Add/Drop Multiplexers: Heuristic and Exact Algorithms, *Operations Research* (to appear) (1998).

Taketal82. K. Takamizawa, T. Nishizeki and N. Saito, Linear Time Computability of Combinatorial Problems on Series Parallel Graphs, *Journal of ACM* **29**, 623–641 (1982).

Thi95. S. Thienel, *ABACUS: A Branch-and-Cut System*, Doctoral Thesis, Department of Computer Science, Universität zu Köln (1995).

Vancetal94. P.H. Vance, C. Barnhart, E.L. Johnson and G.L. Nemhauser, Solving Binary Cutting Stock Problems by Column Generation and Branch-and-Bound, *Computational Optimization and Applications* **3**, 111–130 (1994).

Vdbeck94. F. Vanderbeck, Decomposition and Column Generation for Integer Programs, Ph.D. Thesis, Faculté des Sciences Appliqées, Université Catholique de Louvain, Louvain-la-Neuve (1994).

Vdbeck96. F. Vanderbeck, Computational Study of a Column Generation Algorithm for Bin Packing and Cutting Stock Problems, *Research Papers in Management Studies*, University of Cambridge, no. 1996-14 (1996).

VdbeckWol96. F. Vanderbeck and L.A. Wolsey, An Exact Algorithm for IP Column Generation, *Operations Research Letters* **19**, 151–159 (1996).

Vdbei96. R.J. Vanderbei, *Linear Programming: Foundations and Extensions*, Kluwer, Boston (1996).

VRoWol87. T.J. Van Roy and L.A. Wolsey, Solving Mixed 0-1 Problems by Automatic Reformulation, *Operations Research* **35**, 45–57 (1987).

WagvanHKoe92. A.P.M. Wagelmans, C.P.M. van Hoesel and A.W.J. Kolen, Economic Lot-Sizing: An $O(n \log n)$ Algorithm that Runs in Linear Time in the Wagner-Whitin Case, *Operations Research* **40**, Supplement 1, 145–156 (1992).

WagWhi58. H.M. Wagner and T.M. Whitin, Dynamic Version of the Economic Lot Size Model, *Management Science* **5**, 89–96 (1958).

Wei97. R. Weismantel, On the 0/1 Knapsack Polytope, *Mathematical Programming* **77**, 49–68 (1997).

Wel76. D.J.A. Welsh, *Matroid Theory*, Academic Press, London (1976).

Whi96. Whizzkids '96, Verdien F5000 met een Krantenwijk in Manhattan, De Telegraaf, Technische Universiteit Eindhoven, and CMG (1996).

Wil78. H.P. Williams, *Model Building in Mathematical Programming*, Wiley, Chichester (1978).

Wol75a. L.A. Wolsey, Faces for a Linear Inequality in 0-1 Variables, *Mathematical Programming* **8**, 165–178 (1975).

Wol75b. L.A. Wolsey, Facets and Strong Valid Inequalities for Integer Programs, *Operations Research* **24**, 367–372 (1975).

Wol90. L.A. Wolsey, Valid Inequalities for Mixed Integer Programs with Generalised and Variable Upper Bound Constraints, *Discrete Applied Mathematics* **25**, 251–261 (1990).

Wol97. L.A. Wolsey, MIP Modelling of Changeovers in Production Planning and Scheduling Problems, *European Journal of Operational Research* **99**, 154–165 (1997).

Wri97. S. Wright, *Primal-Dual Interior Point Algorithms*, SIAM Publications, Philadelphia (1997).

Zan66. W.I. Zangwill, A Deterministic Multi-Period Production Scheduling Model with Backlogging, *Management Science* **13**, 105–119 (1966).

Zie95. G.M. Ziegler, *Lectures on Polytopes*, Springer-Verlag, New York (1995).

Index

ABACUS, 161, 241
Affine independence, 142
Algorithm
  primal-dual, 58, 62
  subgradient, 174
Approximation scheme, 217
Aspiration level, 208
Assignment problem, 5, 8, 57, 59, 81
  generalized, 157, 179, 211
  multiplexer, 236, 243
  relaxation, 33
Bc-opt, 161, 241
Benders' reformulation, 201
Bound, 92
  dual, 24, 94, 103, 175
  lower, 238
  primal, 24, 30, 94
  tightening, 104
Branch-and-bound, 95, 160, 223, 239
  active node, 96
  bounding, 95, 99
  branching, 95, 97, 99
    GUB, 102
    strong, 102
  estimations, 99
  LP-based, 98
  node selection, 96–98
  performance, 234
  priorities, 101
  prune
    by bound, 93
    by infeasibility, 94
    by optimality, 93
  reoptimization, 96–98
  system, 101
  tree storage, 98
  updating incumbent, 98
  variable
    fixing, 105–106, 177
    most fractional, 99
Branch-and-cut, 140, 157, 223, 241
Branch-and-price, 193
  See also Column generation
Branching problem, 160
Branching scheme, 193
Changeover, 233, 235
Chinese postman problem, 63
Chvátal-Gomory inequality, 119, 239
Clique, 164
Clustering problem, 195, 198
Column generation, 190, 223, 238–239
  IP, 193
  branching scheme, 197
Combinatorial optimization problem, 4
Complementarity condition, 168
Complexity, 60
Consecutive 1's property, 49
Constraint
  aggregation, 227
  complicating, 167
  joint, 185
  logical, 106

redundant, 104–105
  See also Valid inequality
Convergence, 174
Convex function, 173
Convex hull, 15, 46, 113, 172, 224
  proof, 144
  property, 38, 40, 46, 89
Cooling ratio, 209
Cover, 147
  by nodes, 28, 53
    minimum cardinality, 29
  generalized, 151
  inequality, 147
    extended, 148
    lifted, 159
    separation, 150
Crossover, 210
Cut
  minimum, 43, 49
  pool, 157
Cut-set, 232
  constraints, 8
Cutting plane algorithm, 124, 192, 234, 238
  Gomory fractional, 124
Cutting problem, 3
Cycling, 207
D-inequality, 136
Decomposition, 91, 122, 140, 185
Degree constraint, 170
Delaunay triangulation, 50
Digraph
  acyclic, 68
Dimension, 142
Discrete alternatives, 11
Disjunction, 11, 130
Divide and conquer, 91
Dominance, 141
Dual, 27
  Lagrangian, 168, 174, 192
    solution complexity, 179
    strength, 172, 179
  strong, 43
    property, 38, 40, 89
  variable, 27, 168
Duality, 28, 94
  strong, 28
  superadditive, 33
  weak, 29
Dynamic programming, 67
  recursion, 70
    0-1 knapsack, 73
    integer knapsack, 75–76
    subtree, 72
Easy problem, 37
Efficient
  algorithm, 37
  optimization property, 37, 89, 140

separation property, 40, 89
EMOSL, 241
Enumeration, 8
  implicit, 94
Extreme point, 15, 145
Face, 142
Facet, 142, 226
  proof, 144
Facility location problem, 10, 18
  capacitated, 114
  uncapacitated, 10, 13, 16, 31, 81, 122, 144, 169, 186, 205
Feasible solution, 103
Finite convergence, 133
Fixed charge network flow, 151, 223, 229
Fixed costs, 9
Flow
  conservation constraint, 234
  cover inequality, 152, 224, 229
    separation, 153
  maximum, 43, 81, 155
Forest, 43
Formulation, 5, 9, 12, 223
  better, 16
  dicut, 232
  extended, 14, 227
  ideal, 14
  strength, 232
  strong, 123
  weak, 122
Generalized assignment problem, *see* Assignment problem
Generalized transportation problem, 116
Genetic algorithm, 210
Gomory cut, 127
  mixed integer, 129
Gradient, 174
Graph
  bipartite, 55
  connected, 43
  equipartition, 32, 209
  Eulerian, 212
  series-parallel, 77
  theory, 18
Greedy algorithm, 44, 47
Group problem, 160
Heuristic algorithm, 30, 203, 240
  2-exchange, 206
  cut-and-fix, 216
  dive-and-fix, 214
  greedy, 30, 178, 204–205
  Lagrangian, 177
  local search, 31
  MIP-based, 214
  nearest neighbor, 205
  relax-and-fix, 215, 236
  separation, 156

tree, 213
  tree/matching, 213
  worst case analysis, 211
Incidence matrix, 7
  node-edge, 29
Incumbent, 31
Independent set, 48
Input length, 83
Integer program, 3, 81, 84, 113
  binary, 3, 85
  mixed, 3, 115, 127, 201
  0-1, 114
  totally unimodular, 81
Interior point algorithm, 101
Knapsack problem, 6, 18, 27, 72
  0-1, 6, 9, 12, 16, 30, 73, 81, 84, 86, 114, 147, 150, 180
  integer, 74, 211
Labeling, 55
Lagrange multiplier, 168
Lagrangian, *see* Dual and Relaxation
Lagrangian decomposition, 182
Lift and project, 134
Lifting, 149, 227
Linear programming, 18
  dual, 57
Local minimum, 208
Local optimality, 31
Local search algorithm, 31, 205
  goal function, 205
  neighborhood, 205
  solution, 205
Lot-sizing problem, 11
  backlogging, 77
  capacitated, 85, 227
  constant capacity, 77, 123
  multi-item, 195, 232
  uncapacitated, 11, 17, 68, 81, 83, 147, 223
Master problem, 187
  column generation, 189
  dual bound, 189
  IP, 187, 196
  LP, 188
    restricted, 188
    strength, 192
  optimality condition, 189
  primal feasibility, 189
Matching, 28, 33, 53, 115
  bipartite, 55
  maximum cardinality, 29
  maximum weight, 81
  perfect, 63
Matroid, 48
  intersection algorithm, 62
  optimization problem, 160
MINTO, 161, 241
MIP system, 223, 238, 241

MIPO, 241
MIR inequality, 129, 227
Modeling language, 221, 240
MPS format, 222
Mutation, 210
Neighborhood, 31, 207, 240
Network flow problem, 18
  fixed charge, 69
  minimum cost, 40
Node
  active, 98
  exposed, 54
Odd hole, 135
Optimality conditions, 23, 168
Optimization problem, 89
Partitioning problem, 194
  2-partition, 89
  identical subproblems, 196
Path, 42
  alternating, 53
  augmenting, 54, 56
  longest, 76
  shortest, 42, 67, 81
Performance guarantee, 203
  a priori, 203
Pivoting strategy, 101
Polyhedron, 12
  minimal description, 142
Polynomial
  algorithm, 83, 224
  reduction, 84
Preprocessing, 101, 103
Principle of optimality, 68
Priorities, 234, 239
Problem
  decision, 82
  difficult, 82
  easy, 82
  infeasibility, 105
  legitimate, 82-83
  most difficult, 84
  simplification, 106
  structure, 238
Production planning problem, 2, 18
  See also Lot-sizing problem
Projection, 17, 147
Quadratic 0-1 problem, 27, 155
Reduced cost fixing, 109
Reduction factor, 208
Reduction lemma, 82, 85
Reformulation
  a priori, 121
  automatic, 123
  Dantzig-Wolfe, 187
  multicommodity, 230
Relaxation, 24, 94, 227
  combinatorial, 26

INDEX

Lagrangian, 27, 33, 167, 170, 238
linear programming, 25
state-space, 78
Restriction, 33
Rounding inequality
integer, 116
mixed integer, 117
Routing problem, 77
arc, 90
network, 18
vehicle, 154, 196
Run-time, 235
Running time, 83
Satisfiability problem, 85
Scheduling problem, 1, 18
aircraft, 2
bus, 2
crew, 1
train, 1
Search
best-node first, 100
depth-first, 99
Separation algorithm, 140, 192
Separation problem, 37, 89, 124, 226, 229
cover inequalities, 150
flow cover inequalities, 153
subtour constraints, 155
Set covering problem, 6, 9, 81, 90, 177
Set function, 46
Simplex algorithm, 18, 101
Simulated annealing, 208, 240
Single-source, 229
Software, 221
Stable set problem, 90, 135
Start-up, 77, 233, 235
Subgradient, 174
Submodular, 46
optimization problem, 47
polyhedron, 47
rank function, 48
Subroutine library, 223, 241
Subtour
elimination constraint, 8, 170
elimination constraint
generalized, 154
separation, 155

optimal, 154
Subtree
directed, 231
of a tree, 71
Symmetry, 196, 239
breaking, 239
Tabu
list, 207
search, 207
Telecommunications problem, 2, 77
Temperature, 208
Totally dual integral, 49, 147
Totally unimodular, 38, 41, 49, 145, 155
Tour, 26
2-optimal, 206
Traveling salesman problem, 7, 9, 18, 26, 78, 81, 91
geometric, 175
prize collecting, 154
symmetric, 26, 31, 84, 170, 190, 212
Tree, 27
1-tree, 27, 170
cover, 164
maximum weight, 81
optimal, 43
partitioning, 77
predecessor, 71
root, 71
Steiner, 46, 81
directed, 90
successor, 71
Triangle inequality, 212
Unit commitment problem, 2
Valid inequality, 114, 224
basic mixed integer, 127
blossom, 133
cover, 147
disjunctive, 131
for IP, 118
for LP, 117
for Matching, 118
mixed 0–1, 151
redundant, 141
strong, 139–140, 229
superadditive, 133
Walk, 212

WILEY-INTERSCIENCE SERIES IN DISCRETE MATHEMATICS AND OPTIMIZATION

ADVISORY EDITORS

RONALD L. GRAHAM
AT & T Laboratories, Florham Park, New Jersey, U.S.A.

JAN KAREL LENSTRA
*Department of Mathematics and Computer Science,
Eindhoven University of Technology, Eindhoven, The Netherlands*

ROBERT E. TARJAN
*Princeton University, New Jersey, and
NEC Research Institute, Princeton, New Jersey, U.S.A.*

AARTS AND KORST • Simulated Annealing and Boltzmann Machines: A Stochastic Approach to Combinatorial Optimization and Neural Computing

AARTS AND LENSTRA • Local Search in Combinatorial Optimization

ALON, SPENCER, AND ERDŐS • The Probabilistic Method

ANDERSON AND NASH • Linear Programming in Infinite-Dimensional Spaces: Theory and Application

ASENCOTT • Simulated Annealing: Parallelization Techniques

BARTHÉLEMY AND GUÉNOCHE • Trees and Proximity Representations

BAZARRA, JARVIS, AND SHERALI • Linear Programming and Network Flows

CHONG AND ZAK • An Introduction to Optimization

COFFMAN AND LUEKER • Probabilistic Analysis of Packing and Partitioning Algorithms

COOK, CUNNINGHAM, PULLEYBLANK, AND SCHRIJVER • Combinatorial Optimization

DASKIN • Network and Discrete Location: Modes, Algorithms and Applications

DINITZ AND STINSON • Contemporary Design Theory: A Collection of Surveys

ERICKSON • Introduction to Combinatorics

GLOVER, KLINGHAM, AND PHILLIPS • Network Models in Optimization and Their Practical Problems

GOLSHTEIN AND TRETYAKOV • Modified Lagrangians and Monotone Maps in Optimization

GONDRAN AND MINOUX • Graphs and Algorithms *(Translated by S. Vajdā)*

GRAHAM, ROTHSCHILD, AND SPENCER • Ramsey Theory, Second Edition

GROSS AND TUCKER • Topological Graph Theory

HALL • Combinatorial Theory, Second Edition

JENSEN AND TOFT • Graph Coloring Problems

KAPLAN • Maxima and Minima with Applications: Practical Optimization and Duality

LAWLER, LENSTRA, RINNOOY KAN, AND SHMOYS, Editors • The Traveling Salesman Problem: A Guided Tour of Combinatorial Optimization

LAYWINE AND MULLEN • Discrete Mathematics Using Latin Squares

LEVITIN • Perturbation Theory in Mathematical Programming Applications

MAHMOUD • Evolution of Random Search Trees

MARTELLO AND TOTH • Knapsack Problems: Algorithms and Computer Implementations

McALOON AND TRETKOFF • Optimization and Computational Logic

MINC • Nonnegative Matrices

MINOUX • Mathematical Programming: Theory and Algorithms *(Translated by S. Vajdā)*

MIRCHANDANI AND FRANCIS, Editors • Discrete Location Theory

NEMHAUSER AND WOLSEY • Integer and Combinatorial Optimization

NEMIROVSKY AND YUDIN • Problem Complexity and Method Efficiency in Optimization *(Translated by E. R. Dawson)*

PACH AND AGARWAL • Combinatorial Geometry

PLESS • Introduction to the Theory of Error-Correcting Codes, Third Edition

ROOS AND VIAL • Ph. Theory and Algorithms for Linear Optimization: An Interior Point Approach

SCHEINERMAN AND ULLMAN • Fractional Graph Theory: A Rational Approach to the Theory of Graphs
SCHRIJVER • Theory of Linear and Integer Programming
TOMESCU • Problems in Combinatorics and Graph Theory *(Translated by R. A. Melter)*
TUCKER • Applied Combinatorics, Second Edition
WOLSEY • Integer Programming
YE • Interior Point Algorithms: Theory and Analysis